Environmental Impact Assessments and Mitigation

Environmental Impact Assessments and Mitigation

Musaida Mercy Manyuchi, PhD,
Charles Mbohwa, PhD, Edison Muzenda, PhD,
and Nita Sukdeo, PhD

CRC Press
Taylor & Francis Group
Boca Raton London New York

CRC Press is an imprint of the
Taylor & Francis Group, an **Informa** business

First edition published 2021
by CRC Press
6000 Broken Sound Parkway NW, Suite 300, Boca Raton, FL 33487-2742

and by CRC Press
2 Park Square, Milton Park, Abingdon, Oxon, OX14 4RN

Library of Congress Cataloging-in-Publication Data

Names: Manyuchi, Musaida Mercy, author. | Mbohwa, Charles, author. | Muzenda, Edison, author. | Sukdeo, Nita, author. Title: Environmental impact assessments and mitigation / Musaida Mercy Manyuchi, PhD, Charles Mbohwa, PhD, Edison Muzenda, PhD, Nita Sukdeo, PhD. Description: First edition. | Boca Raton : CRC Press, 2021. | Includes bibliographical references and index. Identifiers: LCCN 2020020227 (print) | LCCN 2020020228 (ebook) | ISBN 9780367220112 (hardback) | ISBN 9780429270307 (ebook) Subjects: LCSH: Environmental impact analysis. Classification: LCC TD194.5 .M36 2021 (print) | LCC TD194.5 (ebook) | DDC 333.71/4--dc23 LC record available at https://lccn.loc.gov/2020020227LC ebook record available at https://lccn.loc.gov/2020020228

ISBN: 978-0-367-22011-2 (hbk)
ISBN: 978-0-429-27030-7 (ebk)

Typeset in Times
by Deanta Global Publishing Services, Chennai, India

Contents

Preface

Environmental issues are a global concern, and measures to preserve the environment must always be prioritized. Developing countries are signatories to the attainment of the United Nations Sustainable Development Goals, which promote development that is conducted in a sustainable manner. Some of the major economic drivers in developing countries include infrastructure development, agriculture, mining and exploration, waste management, and manufacturing. These activities, if not well managed, pose significant dangers to the environment resulting in greenhouse gas emissions and, ultimately, climate change. The need for environmental impact assessments that are conducted professionally and timeously is therefore of utmost importance in order to achieve sustainable environmental management. In this book, environmental impact assessment case studies were drawn from development projects in housing coal bed methane exploration, landfill establishment, small-scale gold mining, and waste metal smelting areas. The projects are described in detail, focusing on how the environment is affected and which mitigation measures can be applied. The authors then propose environmental management plans that align with the underlying environmental management laws of the concerned countries in a bid to promote sustainable development.

About the Authors

Musaida Mercy Manyuchi is a Visiting Associate Professor at the University of Johannesburg in the Faculty of Engineering and the Built Environment. She holds a PhD in Chemical Engineering from the Cape Peninsula University of Technology (South Africa) and MScEng from Stellenbosch University (South Africa). She previously served as Head of Department of Chemical and Process Systems Engineering at Harare Institute of Technology (Zimbabwe) for five years, and has held academic positions in the Department of Operations and Quality Management and Chemical Engineering at the University of Johannesburg, at Manicaland State University of Applied Sciences in Zimbabwe, and at the Biomass Research Institute in Germany. Mercy has worked for more than ten years across various mining sector value chains including cement extraction and processing, coal conversion to value-added products, Environmental Impact Assessments and Environmental Management Plans, research, and capacity-building. Mercy is the Vice Chairperson for Women in Engineering in Zimbabwe. She is also a registered Professional Chemical Engineer with the Engineering Council of Zimbabwe and holds memberships in the World Federation of Engineering Organizations, the World Energy Council, Zimbabwe Institute of Engineers, Professional Women Executives and Business Women's Forum, and the Zimbabwe Institute of Management. Mercy is an alumnus of the prestigious German Green Talents Fellowship and the JF Kapnek Trust Fellowship; a recipient of the Old Mutual Mathematics Olympiad Zimbabwe; and a participant in the Young African Leaders Initiatives Leadership Program.

Charles Mbohwa is the University of Zimbabwe Pro-Vice Chancellor responsible for Strategic Partnerships and Industrialization since 1 July 2019. He is a Visiting Professor at the University of Johannesburg. Before that, he was a Professor of Sustainability Engineering in the Faculty of Engineering and the Built Environment at the University of Johannesburg. Earlier in his career, he was a Mechanical Engineer in the National Railways of Zimbabwe from 1986 to 1991; Lecturer and Senior Lecturer at the University of Zimbabwe, and became Senior Lecturer at the University of Johannesburg in 2007. He was Chairman and Head of Department of Mechanical Engineering at the University of Zimbabwe from 1994 to 1997 and Vice-Dean of Postgraduate Studies Research and Innovation in the Faculty of Engineering and the Built Environment at the University of Johannesburg from July 2014 to June 2017. He was Acting Executive Dean in the Faculty of Engineering and the Built Environment from November 2017 to July 2018. He has published many papers in peer-reviewed journals and conferences. He has published book chapters and several books. He holds a BSc Honours in Mechanical Engineering from the University of Zimbabwe in 1986; Masters of Science in Operations Management and Manufacturing Systems from University of Nottingham; and a Doctor of Engineering from the Tokyo Metropolitan Institute of Technology.

He was a Fulbright Scholar for 6 months in 2006/2007 at the Supply Chain and Logistics Institute at the School of Industrial and Systems Engineering, Georgia

Institute of Technology. He was also a Japan Foundation Fellow. He is a Fellow of the Zimbabwe Academy of Science and of the Zimbabwean Institution of Engineers. He is a registered Mechanical Engineer with the Engineering Council of Zimbabwe. He has been a collaborator in projects of the United Nations Environment Programme. He has been involved in research exchanges with collaborators in many countries including the United Kingdom, Japan, German, France, the USA, Brazil, Sweden, Ghana, Nigeria, Kenya, Tanzania, Malawi, Mauritius, Austria, the Netherlands, Uganda, Namibia, and Australia. In 2016, he was a finalist for the TW Kambule-NSTF Award for research and its outputs by an individual over a period of up to 15 years after award of a PhD, and a finalist in the category Research or Engineering Capacity Development NSTF Awards for achievements over the last 5–10 years. He is a Global Council Member and Fellow of the Industrial Engineering and Operations Management Society. He is a reviewer for more than 20 international journals and is a member of the editorial boards/ committees of 5 journals.

He has collaborated with the United Nations organizations, has assisted in authoring the Global Guidance Principles for Life Cycle Assessment Databases, and has reviewed Methodological Sheets for Subcategories in Social Life Cycle Assessment. He is a founding member of the Social Life Cycle Assessment Alliance coordinated by the World Resources Forum, which focuses on, among many other things, the social impacts of sustainable recycling. He has been a board member of the African Roundtable in Sustainable Consumption and Production. He was also a board member of Almin Industries and Willowvale Motor Industries in Zimbabwe. He has successfully supervised many masters and doctoral students and postdoctoral fellows.

Edison Muzenda is a Full Professor of Chemical and Petroleum Engineering, and Head of Chemical, Materials, and Metallurgical Engineering Department at Botswana International University of Science and Technology. He is also a Visiting Professor in the Department of Chemical Engineering, Faculty of Engineering and Built Environment, University of Johannesburg. He was previously a Full Professor of Chemical Engineering, the Research and Postgraduate Coordinator, as well as Head of the Environmental and Process Systems Engineering and Bioenergy Research Groups at the University of Johannesburg. Professor Muzenda holds a PhD in Chemical Engineering from the University of Birmingham, United Kingdom. His research interests are in green energy engineering, integrated waste management, volatile organic compounds abatement as well as phase equilibrium measurement and computation. He has contributed to more than 280 international peer-reviewed and refereed scientific articles in the form of journals, conferences books, and book chapters. He has supervised more than 30 postgraduate students and over 250 Honours and BTech research students. He serves as reviewer for a number of reputable international conferences and journals. Edison is a member of several academic and scientific organizations including the Institute of Chemical Engineers, UK and South African Institute of Chemical Engineers. He is an editor for a number of scientific journals and conferences. He has organized and chaired several international conferences. He currently serves as an associate editor of the South African Journal of Chemical Engineering. His current research activities are mainly focused on Waste to Energy.

Nita Sukdeo currently works at the Department of Quality and Operations Management, University of Johannesburg. Nita does research in Supply Chain Management and Operations Management, where her current project is "Total Quality Management Practices". She holds a PhD in Engineering Management from the University of Johannesburg.

Abbreviations

AGRITEX	Agricultural Extension Services
BOP	Blowout Preventer
CBM	coal bed methane
D.A	District Administrator
IEA	Initial Environment Audit
EIA	environmental impact assessment
EIS	environmental impact statement
EMA	Environmental Management Agency
EMC	Environmental Monitoring Cell
EMP	environmental management plan
EMPMP	Environmental Management Plan and Monitoring Program
EMS	Environmental Management System
EPFI	Equator Principles Financial Institutions
ERP	Emergency Response Plan
GIS	geographic information systems
g/L	grams per liter
GoZ	Government of Zimbabwe
GSRM	Gas Emission Response Manual
HCC	Harare City Council
HSE	Health, Safety, and Environment
KCl	potassium chloride
LEAP	Local Environmental Action Plan
MENRM	Ministry of Environment and Natural Resources Management
MoHCW	Ministry of Health and Child Welfare
MSW	Municipal Solid Waste
NOSA	National Occupational Safety Association
NSSA	National Social Security Authority
OSRM	Oil Spill Response Manual
OSHA	Occupational Safety and Health Administration
PHG	Prehydrated Gel
PHPA	Partially hydrolyzed Polyacramide
PSC	Production Sharing Contract
ROV	Remotely Operated Vessel
SAIEA	Southern African Institute for Environmental Assessment
SCF	Standard Cubic Feet
SDGs	Sustainable Development Goals
SHE	Safety, Health, and Environment
SI	Statutory Instrument
TDS	Total Dissolved Solids
TOR	Terms of Reference
TSS	Total Suspended Solids
WBM	Water Based Drilling Fluids/Muds

ZERA   Zimbabwe Rural Electrification Agency
ZESA   Zimbabwe Electricity Supply
ZETDC   Zimbabwe Electricity Transmission and Distribution Company
ZINWA   Zimbabwe National Water Authority
ZRP   Zimbabwe Republic Police

1 Background of Environmental Impact Assessment and Management

1.1 INTRODUCTION

The whole world is focused on sustainability, and in Africa, a developing continent, there is much development taking place in terms of infrastructure, mining, manufacturing, and agriculture. These developments, if not carefully managed, could have a negative impact on the environment (World Business Council for Sustainable Development, 2005). Implementation of environmental impact assessments (EIAs) and environmental management plans (EMPs) then becomes critical.

1.2 THE ENVIRONMENTAL IMPACT ASSESSMENT PROCESS

The major steps in the environmental impact assessment (EIA) process are:

  i. Screening
 ii. Prospecting
iii. Scoping
 iv. Full-scale EIA development, review and monitoring

1.2.1 SCREENING

Screening is the process used to determine whether a proposed project requires an EIA and, if so, what level of environmental review is necessary. It also identifies those projects or activities that may cause potential significant impacts (Modak and Biswas, 1999). In addition, it identifies special conditions/analyses that may be required by international funding bodies before conducting the full EIA. The screening process further categorizes the projects to indicate where:

  i. A full-scale EIA is required
 ii. Some further environmental analysis is required
iii. No further environmental analysis is required

Examples of projects that need full EIAs include but are not limited to:

    i. Infrastructure projects
    ii. Large-scale industrial activities
    iii. Resource extractive industries and activities
    iv. Waste management and disposal
    v. Substantial changes in farming or fishing practices

The initial screening criteria typically consider:

    i. Project type, location, size (e.g. capital investment, number of people affected, project capacity, areal extent)
    ii. Receiving environment characteristics
    iii. Strength of community opinion
    iv. Confidence in prediction of impacts

The project screening checklist should include a section regarding site location, characteristics, including, at a minimum, the four categories of environmentally critical areas also classified as highly sensitive areas:

    i. National parks
    ii. Indigenous people's areas
    iii. Tourist areas
    iv. Ecologically sensitive areas

Site selection defines the location of the study area and the specific environmental resource base to be examined. The most important factor is to look at all components contributing to a project's potential negative impacts (Ahmad and Sammy, 1987). Existing regional development plans should be used as guides to select project locations where environmental conditions will be minimally impacted.

1.2.2 PROSPECTUS GENERATION AND IMPACT IDENTIFICATION

Initial prospectus document generation is intended as a low-cost environmental evaluation that makes use of information already available. The prospectus describes the proposed project or activity, examines alternatives, identifies and addresses community concerns to the fullest possible extent, identifies and assesses potential environmental effects, thereby directing future actions of the proposed project (Lee, 1995); and identifies all potential environmental concerns relating to a proposed project or activity, thereby establishing a focus for follow-up studies for the comprehensive EIA. The initiation of the prospectus document results in the following project classifications:

    i. No requirement for further environmental study; proposal not anticipated to have significant impact
    ii. Limited environmental study needed; environmental impacts are known and can be easily mitigated
    iii. Full-scale EIA required; impacts unknown or likely to be significant

In order to establish the potential significant issues for any project in question, it is important to first identify valued environmental/ecosystem components. This can be done through hiring professional consultants, reviewing past experience and legislative requirements, assessing stakeholder and community values, as well as practical identification of the potential impacts of the project. The prospecting stage also identifies the potential for cumulative impacts (i.e. to the site as a whole and to the region). The commonly considered valued environmental ecosystems are:

    i. Natural physical resources (e.g. surface and groundwater, air, climate, and soil)
    ii. Natural biological resources (e.g. forests, wetlands, river, and lake ecology)
    iii. Economic development resources (e.g. agriculture, industry, infrastructure, and tourism)
    iv. Quality of life (e.g. public health, socioeconomic, cultural, and aesthetics)
    v. National commitments (e.g. endangered species protection)

The potential impacts during the prospecting stage can be identified using matrices (sectorial or project type), checklists, professional expertise, past experience with similar type projects, or a combination of techniques. Various information that is required during the prospectus development includes the project type, size and location, the areas of potential impact (such as physical, biological, and economic development resources, as well as quality of life and other existing or planned projects).

    The prospectus document can be developed using various sources; information such as existing reports on environmental resources in the area, previous assessment reports, prospectuses, and EIA reports generated on similar type projects, reports on other projects in the region that may cause similar disturbances; regional planning, policy, and other reports; field studies; and local citizens' and traditional knowledge can be used. The impacts from the proposed project can be classified based on their effects as follows:

    i. Nature: positive, negative, direct, indirect, cumulative, synergistic
    ii. Magnitude
    iii. Extent/location: area/volume covered, distribution
    iv. Timing: during construction, operation, decommissioning, immediate, delayed, rate of change
    v. Duration: short-term, long-term, intermittent, continuous
    vi. Reversibility/irreversibility
    vii. Likelihood: risk, uncertainty, or confidence in the predictions

The criteria for evaluating environmental effects can be based on importance of affected resource, magnitude and extent of disturbance, duration and frequency, risk/likelihood of occurrence, reversibility and contribution to cumulative impacts. In summary, the details in a prospectus report which will feed into the EIA will contain the following:

    i. Description of the project
    ii. Description of the environment

 iii. Screening of potential environmental issues and rationale for their significance grading
 iv. Environmental protection measures
  v. Environmental monitoring and institutional requirements
 vi. Recommendations for further studies
 vii Conclusion

1.2.3 SCOPING

Scoping is a process of interaction between the consultant, government agencies, other relevant stakeholders, and project proponents. During the scoping phase, the following items are identified:

  i. Spatial and temporal boundaries for the EIA
  ii. Important project issues and concerns
 iii. Information necessary for decision-making
 iv. Significant effects and factors to be considered
  v. Terms of reference for full-scale EIA development

Project scoping is critical as it serves to facilitate an efficient EIA by identifying appropriate areas for consideration (e.g. key issues, concerns, alternatives). Furthermore, it reduces the likelihood of deficiencies in the EIA, ensuring that important issues are not overlooked. Scoping also prevents unnecessary expenditures and time delays from oversights or unnecessary areas of study. During scoping, the background information of the project is given and this normally comprises the project description (i.e. type, magnitude, location, alternatives, and constraints), environmental setting (i.e., delineation of study area and list of environmental resources and sensitive areas) and background reports (e.g. aspects of the environmental setting and previous projects with relevant impacts or resources).

1.2.4 FULL-SCALE EIA DEVELOPMENT, REVIEW, AND MONITORING

The EIA will comprise of all the items listed below for the project under review. The impacts and their significance are assessed, using both qualitative and quantitative methods. The summary of the EIA process is shown in Figure 1.1.

 The mitigation development phase and the appropriate mitigation measures are then recommended. The impacts during the EIA development can be identified through the following qualitative and quantitative methods: checklists, matrices, networks, geographic information systems (GIS), expert systems, and risk assessment. The advantages and disadvantages of the various methods are shown in Table 1.1. The specific EIA requirements typically include the following:

  i. EIA objectives
  ii. Legal and governmental policy requirements
 iii. Significant environmental issues of concern

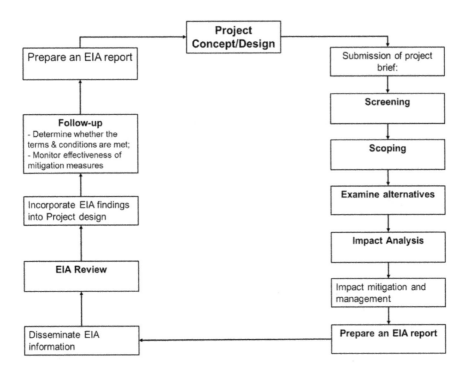

FIGURE 1.1 The EIA process algorithm.

TABLE 1.1
Methods Used for Determining Impacts during EIAs Development

Method	Advantages	Disadvantages
Checklists	Simple to understand and use Good for site selection and priority-setting	Do not distinguish between direct and indirect impacts Do not link action and impact Qualitative

Matrices	Link action to impact
	Good method for displaying EIA results
		Difficult to distinguish direct and
		  indirect impacts
		Significant potential for double-
		  counting of impacts

Networks	Link action to impact
	Useful in simplified form in checking for
	  second-order impacts
	Handles direct and indirect impacts
		Can become overly complex if used
		  beyond simplified version
		Qualitative

Overlays	Easy to understand and use Good display method Good for site selection	Addresses only direct impacts Does not address impact duration or probability
Expert systems	Excellent for impact identification and analysis Good for experimenting Semi-quantitative to quantitative	Heavy reliance on knowledge and data Often complex and expensive

iv. Required information and data methodologies for impact assessment
v. Process for incorporating public input through consultations
vi. Impacts identification, their significance, and mitigation
vii. Environmental management plan for all the phases

The impact determination method is based on the type and size of the proposal, type of alternatives being assessed, nature of likely impacts, experience using EIA methods, and the resources available, as well as the nature of public involvement and the procedural/administrative requirements. The significance of the environmental impact is determined as below.

$$ImpactCharacteristics \left(spatial\ extent\right) \times ImpactImportance \left(e.g.\ value\right)$$

$$= ImpactSignificance$$

The characteristics affecting impact significance include the nature of the impact (e.g. positive, negative, synergistic), the extent and magnitude, timing (i.e. construction operation and closure), and duration (i.e. short, chronic, intermittent). Factors such as reversibility/irreversibility of the impact and likelihood (i.e. probability, uncertainty) of it occurring are also important. Significance criteria include:

i. Importance: the value that is attached to the affected environmental component
ii. Extent of disturbance: the area expected to be impacted
iii. Duration and frequency of disturbance
iv. Reversibility
v. Risk: probability of an unplanned incident caused by the project

The significance of each impact is based on considerable expert judgment and technical knowledge, which are often required to fully understand the nature and extent of environmental impacts. The categories for significance include no impact, unknown impact, significant impact, mitigated impact, and insignificant impact. For assessment of the significance of the impacts it is important to:

i. Use rational and objective methods
ii. Provide consistency for comparison of project alternatives
iii. Document values and beliefs used in making judgment decisions
iv. Apply the following impact significance criteria:
   a. Ecological importance/sustainability criteria
   b. Social importance
   c. Environmental standards

1.3 APPLICATION OF EIA TECHNIQUES IN KEY DEVELOPMENT PROJECTS

African countries are implementing the 17 Sustainable Development Goals (SDGs) in the various economic areas. For the SDGs to be completely realized, environmental management is very important (ISO14040, 2006). In this work, various developmental projects from infrastructure, mining, logistics, manufacturing, recycling, and social services were considered, and the EIAs and EMPs were developed for implementation, depending on the nature of the project.

1.4 CONCLUSION AND RECOMMENDATIONS

1.4.1 CONCLUSION

EIAs provide the means by which to make informed decisions regarding development of the environment. As such, their implementation is of paramount importance in the effort to promote sustainable development and preserve the environment for future generations. Although every development project comes with negative impacts to the environment, EIAs help to strike a balance between negative impact mitigation and positive impact maximization.

1.4.2 RECOMMENDATIONS

EIAs must be done for all key projects. This is crucial for promoting green and sustainable development that will result in responsible management of the environment in developing countries.

REFERENCES

Ahmad, Y. J. and Sammy, G. K. (1987). Guidelines to Environmental Impact Assessment in Developing Countries. *UNEP Regional Seas Reports and Studies No. 85*, UNEP.

ISO14040 (2006). International Organization for Standardizations: Environmental Management - Life Cycle Assessment-Principles and Frameworks. Geneva, Switzerland. [5] DIN 53183 (1973) Paints, Varnishes and Similar Products.

Lee, N. (1995). Environmental Assessment in European Union: A Tenth Anniversary Project Appraisal 7, 123–136.

Modak, P. and Biswas, A. K. (1999). *Conducting Environmental Impact Assessment for Developing Countries*. Tokyo; New York: United Nations University Press.

World Business Council for Sustainable Development (2005). Environmental and Social Impact Assessment (ESIA) Guidelines, 54 pp.

2 Environmental Impact Assessments and New Residential Infrastructure

2.1 BACKGROUND

Urban populations in Zimbabwe continue to grow at an alarming rate, while local urban housing delivery systems lag woefully behind. Although real estate has been used up, population density is high, and municipal service delivery is at risk, there is still a great unmet need for housing. Currently, the demand for residential stands (**residential** area contains houses rather than offices or factories) in Zimbabwe is greater than 500 000 units, with the largest shortages in major cities where the level of unemployment has also risen tremendously. This demand for residential stands is characterized by a growing number of housing cooperatives that work to find desirable residential stands for their members.

Because most local authorities lack the resources needed to provide standard urban accommodations, the national government intervened by introducing the National Housing Delivery Program to assist in housing provision. Under the program, the government facilitates home ownership for its citizens through acquisition of land for urban development, while the beneficiaries are ultimately responsible for building their own homes using their own or borrowed funds.

A layout plan of the residential site must be prepared and approved by the Director of Physical Planning in terms of the Regional Town and Country Planning Act, Chapter 29:12. The layout plan provides for low-density residential stands, commercial stands, and recreational stands.

The shortage of residential stands, and consequently, appropriate urban housing, is one of a multiplicity of social and economic challenges that are symptomatic of Zimbabwe's unstable, unpredictable, and insecure macroeconomic environment. Others include a huge national budget deficit, high interest and unemployment rates, very high levels of poverty, and low and continuously declining per capita income.

These factors not only make residential and commercial stands unavailable, inaccessible or unaffordable, they also impede the capacities of local authorities, companies, and individuals to effectively deliver industrial and residential stands.

The intent of this proposed project is to construct green residential stands, thereby reducing the backlog, and to generate jobs during the construction phase. After construction, not only will further employment opportunities be created, in the areas of mechanical and commercial services, but local revenue streams are also expected by the infrastructure proponents to increase with the newly added infrastructure service charges.

This section provides a broad description of the infrastructure area and the construction work associated with it.

2.2   SCOPE, OBJECTIVES, AND CRITERIA OF THE HOUSING EIA REPORT

2.2.1   SCOPE OF REPORT

The Environmental Management Agency (EMA) policy for all new infrastructure, programs, and activities requires that an environmental impact assessment be carried out during the planning stage to ensure that significant impacts on the environment are taken into consideration during the design, construction, operation, and decommissioning of the facility.

The environmental impact assessment (EIA) report focuses on:

    i. The baseline environmental conditions of the area
   ii. Description of the proposed infrastructure
  iii. Provisions of the relevant environmental laws and legislations
   iv. Public consultation
    v. Identification and discussion of any adverse impacts to the environment anticipated from the proposed infrastructure
   vi. Appropriate mitigation measures
  vii. Provision of an environmental management plan outline (United Nations, 2000; REMA, 2005)

2.2.2   TERMS OF REFERENCE FOR THE EIA PROCESS

The terms of reference for the EIA are:

    i. To describe the proposed infrastructure and associated works together with the requirements for carrying out the proposed development
   ii. A review of the national environmental legislative and regulatory framework, baseline information, and any other relevant information related to the infrastructure
  iii. Analysis of alternatives including infrastructure site, design, and technologies and the reasons for preferring the proposed site design and technologies
   iv. To identify and describe the elements of the community and environment, at large
    v. Neighbors likely to be affected by the proposed developments
   vi. To establish the baseline environmental and social scenario of the infrastructure site and its surroundings
  vii. To identify emission sources and determine the significance of impacts
 viii. To identify, predict, and evaluate environmental and social impacts during the construction and implementation of the infrastructure, in relation to the environmental variables
   ix. To develop an environmental management plan (EMP) that identifies negative impacts and proposes mitigation measures so as to minimize pollution, environmental disturbance, and nuisance during construction and operations of the development

x. To design and specify the monitoring/audit requirements necessary to ensure the implementation and the effectiveness of the mitigation measures that are adopted

The output from the consultants includes:

i. An EIA study report comprised of an executive
1. summary, study approach, baseline conditions, anticipated impacts, and proposed mitigation
ii. An environmental management plan (EMP) outline. This EMP forms part of the report recommendations

2.3 INFRASTRUCTURE AREA

The infrastructure area for the residential development must be given. The current plans in accordance with the city council must be disclosed. The infrastructure site must then be indicated on a 1:50 000 map.

2.4 GENERAL DESCRIPTION OF THE INFRASTRUCTURE DEVELOPMENT

The design of the layout has taken into consideration the basic principles of economy, efficiency, and convenience by proposing the best possible interrelationships among various land uses. Open spaces have been provided so as to create a pleasant living environment, by acting as natural "breathers", and act as buffers to street traffic. Incidental open space, such as non-developable (or sloppy) ground, has also been included.

The provisions of the layout plan include residential stands, commercial stands, and recreational stands.

2.4.1 MAJOR INFRASTRUCTURE

A number of utilities will be established on-site in order to fully service the stands. The major activities at the site will include the following:

Road construction: Will be conducted following the provisions of the layout plan as well as the road designs, all approved by the city council

Water reticulation: The proponent proposed that tenants will be provided with water service by the city council.

Sewer reticulation: Due to the long distance from the town center to the infrastructure site, coupled by the scarcity of water, a soak away system will be used for sewage management.

Solid waste management: The area will be serviced by the City Council. The predominant waste will be household communal waste. The waste includes plastics, cardboard boxes, tins, bottles, and organic waste.

Electricity lines: Plans to introduce power line systems into the area must be completed. The infrastructure will be connected to the Zimbabwe Electricity Supply Authority (ZESA) Power supply lines shall be installed and fed to each unit by ZESA at the expense of the infrastructure proponent, and as specified by the engineers.

Drainage utilities: The land expanse is adjacent to a range of mountains, and rain water is expected to flow toward the residential area along waterways or areas of low depression. This water will eventually collect in the underground water system. To guard against flooding of the area, 350 mm diameter pipe drainage systems will be installed to collect surface water. Shallow roadside concrete slab canals and underground tunnels will be constructed to collect all the water from various areas of the stands, in accordance with agreed upon road design standards.

Accommodation for site workers: Out-of-town site workers will be accommodated in existing shelters at the infrastructure site. The shelters have toilet and bathing facilities as well as tapped water. Local workers will stay at their homes and report for work daily.

2.5   ENVIRONMENTAL ASSESSMENT AND MONITORING PLAN

An EIA study of the proposed infrastructure and documentation of the study needs to be completed.

2.5.1   ENVIRONMENTAL ASSESSMENT OBJECTIVES

The objectives of the housing EIA assessment include the following:

  i. Identify the baseline conditions of the infrastructure area
 ii. Predict or identify potential negative environmental impacts and their associated effects during and after infrastructure implementation
iii. Develop or suggest measures for avoiding or mitigating the predicted negative impacts
 iv. Incorporate environmental considerations in the infrastructure design and operation so that the infrastructure becomes sustainable and environmentally sound

In considering the effects of the proposed residential infrastructure, it is necessary to recognize the potential environmental impacts, both positive and negative, on the surrounding area and infrastructure area during the periods of construction, operation, and maintenance.

2.5.2   ENVIRONMENTAL ASSESSMENT METHODOLOGY

The assessment must be carried out in four stages, namely:

  i. Desktop study planning
 ii. Public consultation

iii. Baseline survey
iv. On-site investigations

2.5.2.1   Desktop study and planning

The scoping of the infrastructure, in which the breadth, depth, and overall approach of the EIA are established prior to beginning detailed investigations. Scoping serves to focus the EIA early in order to ensure that environmental issues and impacts of greatest significance receive the most attention during the investigation and analysis phases. As part of the scoping process, stakeholders and interested parties in the infrastructure are identified. The scoping process is done to ensure that all aspects of the infrastructure are considered and factored into the EIA process. The information gathered during the scoping exercise is used to guide the rest of the EIA process.

2.5.2.2   Public consultation

Public consultation is carried out in order to:

i. Inform the various stakeholders about the proposed development activities and obtain their views
ii. Evaluate the potential environmental impacts of the proposed development infrastructure
iii. Assess the need of such an infrastructure and how the establishment of the infrastructure could solve the current residential stands backlog in most city councils
iv. Find areas of potential conflict or complementarities with other development infrastructures in the area
v. Consult with the stakeholders through interviews; questions are tailored for each group of respondents

2.5.2.3   Baseline survey

A baseline survey, which includes a desk study, is conducted in order to establish a profile of the existing environment, against which an assessment of the various impacts brought about by the planning, construction, and operational phases of the infrastructure can be made. The baseline data collected include the following:

i. Geology and hydrology of the study area
ii. Inventory and distribution of flora and fauna
iii. Soil types and characteristics
iv. Climate (rainfall and temperature) data
v. Land use pattern in the study area
vi. Prevailing socioeconomic conditions
vii. Development plans for the area (EPA, 2008).

2.5.2.4   On-site investigations

The final step in the assessment process is an on-site investigation in which the EIA team visits the infrastructure site to:

i. Fill gaps in the gathered baseline information

ii. Ensure that potential impacts identified in the scoping exercise as being most significant for the infrastructure were adequately addressed

iii. Predict and assess the significance of environmental impacts, including those during construction and operational phases of the infrastructure

iv. Gather detailed information needed to define, describe, and estimate costs of any mitigation measures to be proposed

v. Evaluate institutional requirements and capabilities for effective impact mitigation, enhancement of positive impacts, and environmental monitoring

vi. Develop a plan to enable the ongoing monitoring and evaluation of the infrastructure during both the construction and operation phases

The information gathered during the on-site investigations, in conjunction with data collected in the earlier phases, forms the basis for the environmental impact analysis and the suggested mitigation and enhancement measures.

2.6   JUSTIFICATION OF THE INFRASTRUCTURE DEVELOPMENT

Failure to adequately address the backlog of residential and commercial stands will jeopardize economic growth. With this in mind, the Zimbabwe government recently adopted a policy allowing that municipalities and councils be assisted by individuals, companies, and others, such as institutions, to buy land for development – either for industry or housing purposes. It is appreciable and pleasing to note, therefore, that the proponent of this infrastructure project has embarked on this necessary and important effort which will significantly contribute to the Zimbabwe gross domestic product as well as to that of Matabeleland South Province. However, it is crucial that the infrastructure should not be implemented at the expense of the environment; hence the need to carry out an environmental impact assessment before the infrastructure commences, in line with EIA regulations as enshrined in the Environmental Management Act.

2.7   POLICY AND LEGAL ADMINISTRATIVE FRAMEWORKS

2.7.1   INTRODUCTION

Any proposed infrastructure will be subject to a number of pieces of legislation. The following statutes, policies, and regulations are important for the successful implementation of the infrastructure.

2.7.2   ENVIRONMENTAL IMPACT ASSESSMENT POLICY

The Interim Environmental Impact Assessment (EIA) Policy was published in July 1994. This policy has been amended after taking into account submissions from stakeholders and the experience of the Ministry of Environment and Tourism in implementing this policy. A final policy document was then published in August 1997. The goals of the EIA policy are:

i. To encourage environmentally responsible investment and development in Zimbabwe

ii. To maintain the long-term ability of natural resources to support human, plant, and animal life
iii. To conserve a broad diversity of plants, animals, ecosystem, and the natural processes that they depend upon
iv. To conserve the social, historical, and cultural values of people and their communities

The EIA Policy has 9 principles some of which are:

i. Sustainability for future generations is the cornerstone of environmental management
ii. EIA must enhance development by contributing to its environmental sustainability, not inhibit it
iii. EIA is a means for infrastructure planning, not just evaluation.
iv. Infrastructure impacts must be monitored throughout the life of the development
v. Particular attention must be given to the distribution of infrastructure cost and benefits

The Policy requires that EIA studies be undertaken for all new infrastructures. Infrastructures involving construction on a large scale are prescribed for EIA.

The Policy also requires that the public, particularly the stakeholders, be consulted during the EIA study.

2.8   ENVIRONMENTAL MANAGEMENT ACT (20:27) OF 2003

The Act seeks to harmonize all pieces of legislation governing the environment, encompassing both brown (water, waste, and pollution) and green issues. Environmental impact assessment is an integral part of the Act and is now compulsory. The objectives of this Act are to provide for the sustainable management of natural resources, protection of the environment, and prevention of pollution and environmental degradation.

It thus embraces all the functions of the Natural Resources Act, the Atmospheric Pollution Prevention Act, the Hazardous Substances and Articles Act, and the Noxious Weeds Act. It has also taken over the section of the Water Act that were dealing with water pollution and the sections of the Forestry Act dealing with natural resources. In Section 3(2), the Act, states that if any other law is in conflict or inconsistent with it, then the Environmental Management Act shall prevail.

Major construction works are considered infrastructures, and must not be implemented unless an EIA has been conducted (First Schedule).

The sections relevant to the EIA are Sections 97 to 108 of Part XI:

Section 97: Infrastructures for which environmental impact assessment is required
This section outlines the infrastructures that require an EIA and makes it an offense to carry out such infrastructures without the necessary approvals.

Section 98: Developer to submit prospectus

This section states that before an EIA is conducted the developer must submit a prospectus to the Director General. The prospectus must contain such information regarding the assessment and the infrastructure as may be prescribed. The Director General then examines the prospectus and can either approve or reject the prospectus.

Section 99: Contents of environmental impact assessment report

This section lays down the contents of an EIA report which include a detailed description of the infrastructure and its activities; reasons for selecting proposed site; detailed description and likely impacts; and the specification of the measures for eliminating, reducing, or mitigating the likely impacts.

Section 100: Consideration of environmental impact assessment report and issue of certificate

This section specifies that the Director General shall consider the EIA report within sixty days from the date of receiving the report. If he/she fails to notify the developer within that period, the infrastructure shall be deemed to have been approved. The infrastructure may be approved or rejected. Upon approval of an infrastructure, the Director General shall issue a certificate to the developer.

Sections 101 to 105: Certificates

The certificates issued shall be for a period of two years and be extended if the infrastructure has not been completed. If however the infrastructure has not commenced by the end of the time period, the certificate will not be extended. A register of these certificates will be maintained, and the certificates can be amended, suspended or even cancelled if it is realized that the infrastructure will indeed be a source of pollution. The developer may notify the Director General if the infrastructure fails to take off as scheduled.

Sections 106-108: Environmental audits and inspections

Periodic audits of any infrastructure will be carried out to ensure that the infrastructure implementation complies with the requirements of the Environmental Management Act.

The developers shall also take all reasonable steps to prevent or mitigate any adverse effects arising from the infrastructure and shall keep the Director General informed. The EIA report shall always be open for public inspection at the Director General's office. The Environment Management Act was made operational on the 17th of March 2003. The developer will have to satisfy all the provisions of the Act including the preparation of an EIA report.

2.9   PARKS AND WILDLIFE ACT

The Parks and Wildlife Act provides for the preservation, conservation, propagation, and control of wildlife, birds, fish, and plants of Zimbabwe and the protection of the country's natural landscape and scenery. Therefore, requirements of the Act must be considered and followed during the construction and existence of residential housing infrastructure. The construction team must be instructed not to engage in activities which will violate this Act. The following is a summary of the relevant sections as pertaining to construction work.

Section 24:
Provides for the protection of plants, animals, or any wildlife that may be in danger from hunting or gathering.
Section 28:
Forbids anyone from picking plants in botanical garden areas.
Sections 33 and 38:
Controls the hunting and removal of animals from game sanctuaries and safari areas.
Part IX and Part X:
Protects specially protected animals and plants from being hunted, picked, or sold.
Part XI:
Protects indigenous plants from being picked unless permission has been granted.
Part XII:

Controls the hunting and removal of live animals and animal products. It outlines when and how authority can be obtained to do so. The developer will follow all the provisions in the Act. The developer will have to work with the Department of Parks on implementation to ensure that animals are protected.

2.10   ROAD TRAFFIC ACT

The Road Traffic Act proscribes construction that encroaches upon a restricted road. Part of Section 38 of the Act states:

> Where any structure or works for which no authority has been granted under section thirty-six erected or carried or any structure or work is erected or carried out in contravention of any condition imposed under any other enactment, the Minister may by notice in writing direct to the owner or person having control or possession thereof to remove it or to make such alterations thereto as may be specified in such notice and to carry out such removal or alterations within such period, which shall not be less than thirty days as from the date of such notice, as shall be specified in such notice.

Servitudes to the road were granted at planning stage and developer will follow the road designs as provided in the layout plan.
(Statute Law of Zimbabwe, Volume 2 revised 1996)

2.11   REGIONAL, TOWN, AND COUNTRY
PLANNING ACT CHAPTER 29:12

This act provides for the planning of regions, districts, and local areas with the objective of conserving and improving the physical environmental and, in particular, promoting health, safety, amenity, convenience, and general welfare. The Regional, Town, and Country Planning Act allows for the formulation of regional plans, local plan, and a master plan. A regional plan covers areas under several local authorities. It indicates the major land uses, including the important public utilities, and major amenity and recreational areas; areas for development; major transportation and communication patterns; and measures for the conservation and improvement of the physical environment.

Regional plans are formulated by regional planning councils that comprise representatives of local authorities and the major activities within that region. A master plan is a broad plan for the whole local authority. It will show areas planned for various land uses within the local authority. The act allows for the formulation of regional, district, master, and local plans. Of particular interest to this study is the local plan because the subject property falls under one such plan.

Local plans are prepared according to Section 17 of the Act. Section 17 (3) (ii) states that the purpose of a local plan is to formulate proposals for the coordinated and harmonious development of an area, including measures for the conservation and improvement of the physical environment. Stand 8363 to 8406 wholly falls under the town of Harare local plan, whose provisions are covered in detail in the section that deals with the land use issue.

Any person wishing to subdivide property must apply to the local planning Authority for permission according to Sections 39 and 40 of the Act. Section 39 (1) states that "no person shall subdivide any property without a subdivision permit". Section 40 (1) states that "an application for a permit to subdivide/consolidate property shall be made to the local planning authority in such a manner as may be prescribed and shall be accompanied:

  i. By the consent in writing of the owner of the property and every holder of a mortgage bond registered over the property
  ii. By the consent in writing of the holder of any other real right registered over the property, if required by the local planning authority

The local planning authority is obliged, under Section 40 (2), to acknowledge receipt of the application within two weeks of receiving such an application unless the application is incomplete, in which case it shall acknowledge receipt thereof as soon as the application is complete. A town planning subdivision and consolidation (TPSC) form duly completed accompanies the application.

Sub-section 40 (3) (a) provides for the procedures that must be adopted by the local authority when dealing with a proposal which conflicts with any condition which is registered against the title deeds of the property.

The infrastructure under discussion here has been approved by The Director of Physical Planning according to Section 43 of the Regional Town and Country Planning Act Chapter 29:12. The provisions of the layout plan are legally binding and have to be strictly adhered to. The developer will ensure that no amendments are made to an approved layout plan without the consent of the Ministry of Local Government through the Director of Physical Planning.

2.12   PUBLIC HEALTH ACT

The relevant aspects of this act are those dealing with sanitation and housing at the site and those temporary structures which will be applicable when accommodating workers during the construction of the residential stands. The Act prohibits creation of nuisance, and the relevant definitions of nuisance given are:

i. Any dwelling or premises which is of such construction or in such a state or so dirty or so verminous as to be injurious or pose a danger of infectious disease

ii. Any sanitary convenience that is foul or in such a state or situated or constructed as to be offensive or to be injurious or dangerous to health; or any collection of water which may serve as a breeding pool for mosquitoes

iii. Any well or other source of water supply, whether public or private ... the water of which is used or likely to be used by man ... which is polluted or otherwise liable to render any such water injurious or dangerous to health

iv. Any accumulation or deposit or refuse or other matter whatsoever which is offensive or which is injurious or dangerous to health

v. Any dwelling which is so overcrowded as to be injurious or dangerous to the health of the residents

During the construction of the infrastructure, contractors will need to be made aware of the need to abide by this legislation when providing accommodation for the labor force so that they will meet the requirements of this Act. Of particular concern is the provision of waste and sewage-handling facilities to match the population anticipated.

2.13   EXPLOSIVES ACT (10.08)

Regulations provided under this Act describe the blasting and safety procedures that are required by law to be observed during the use of explosives. A blasting schedule has to be submitted to the Inspector of Mines, and regulations require that this schedule be displayed publicly.

The Inspector of Mines shall describe the method of blasting in cases where this activity is potentially dangerous to people, buildings, and roads. Before blasting, a warning siren is required to be sounded. This Act is pertinent to those sections where blasting may be required. However, no blasting is anticipated in the current infrastructure.

2.14   FORESTRY ACT

According to Section 38: Reservation of trees or forest produce of the Forestry Act:

The Minister may, in respect of any state forest, by statutory instrument, declare any species of tree or any forest produce to be especially reserved and may in like manner revoke, or amend, any such declaration.

Section 39: Protection of forest or trees from cutting stipulates:

Whenever, in respect of any land not being a state forest, the President deems it expedient in the public interest that any tree or the whole or any part of a forest or plantation shall be protected, the President may, by proclamation, declare such tree or such forest part of a forest or plantation to be protected.

The owner of any forest, plantation, or tree of which the President has exercised his power under Subsection (91) shall be entitled to compensation for any loss resulting herefrom in such as may be mutually agreed upon or, failing agreement, as may be determined by arbitration.

Minimum clearance of trees shall be done at implementation and the developer shall ensure that no wood is transported without permission from the Forestry Commission in Zimbabwe.

2.15   PNEUMOCONIOSIS ACT [CAP 15:09]

This Act provides for the control and administration of persons in occupations where dust inhalation is an issue. A medical bureau must be established under this Act to *inter alia* perform medical assessment and record incidences of pneumoconiosis. Workers are issued certificates that certify fitness to work in dust conditions.

Site workers shall be provided with protective clothing to avoid inhaling dust.

2.16   ATMOSPHERIC POLLUTION ACT (1971)

Under the Act, uncontrolled production and disposal of artificial waste are prohibited. The Act provides for the prevention and control of air pollution by noxious or offensive gases, dust, smoke, and internal combustion fumes. Although this act was repealed, the ground shall still be dampened at the implementation stage to avoid dust. Construction workers will wear protective clothing to avoid infection.

2.17   HAZARDOUS SUBSTANCES AND ARTICLES CONTROL ACT (1971)

This Act regulates dumping of industrial poisons, pesticides, and other dangerous substances and waste, including radioactive materials. Although this act was repealed, the developer will ensure that no hazardous substances are dumped in the infrastructure area or at any site not suitable for hazardous waste.

2.18   WATER ACT 20:24 (1998)

Part IV of the Water Act makes provision for the control of water pollution and protection of water resources. In Sections 67 to 71 of the Act, provision is made ensuring that water resources management is consistent with broader National Plan CAP 20:27 SI No 2007. The maximum permissible concentration of chemical constituents permissible in water which is discharged or disposed of in a Zone 1 or Zone 2 catchment shall be specified as shown in Table 2.1.

TABLE 2.1
Maximum Permissible Concentration of Selected
Chemicals in mg/L

Chemical constituent	Zone 1*	Zone 2**
Cadmium (Cd)	0.01	0.01
Chromium (Cr)	0.05	0.05
Cyanide and other related compounds	0.2	0.2
Mercury (Hg)	0.5	0.5
Nickel (Ni)	0.3	0.3
Zinc (Zn)	0.3	0.3
Iron (Fe)	0.3	0.3
Total heavy metals	1.0	2.0

* Zone 1: Catchment areas in Zimbabwe's Agro-ecological Region I
** Zone 2: Catchment areas in Zimbabwe's Agro-ecological Region II–IV

In addition, the water should not contain any detectable quantities of pesticides, herbicides, or insecticide or any other substance not referred to elsewhere in these standards in concentrations that are poisonous or injurious to human, animal and vegetable, or aquatic life.

In line with this legislation, the developer will have to apply for a permit from the Zimbabwe National Water Authority (ZINWA) prior to establishing supplementary water sources such as boreholes. Owners of individual stands will also need to apply for a permit from ZINWA to sink a borehole at their premises.

2.19 LABOR ACT [CHAPTER 28.01]

This Act declares and defines the fundamental rights of employees, in adherence to the international obligations of the Republic of Zimbabwe as a member state of the International Labour Organization and as a member of or party to any other international organization or agreement governing conditions of employment which Zimbabwe would have ratified; to define unfair labor practices; to regulate conditions of employment and other related matters; to provide for the control of wages and salaries; to provide for the appointment and functions of workers committees; to provide for the formation, registration, and functions of trade unions, employers organizations and employment councils; to regulate the negotiation, scope and enforcement of collective bargaining agreements; to provide for the establishment and functions of the Labour Court; to provide for the prevention of trade disputes, and unfair labor practices; to regulate and control collective action; to regulate and control employment agencies; and to provide for matters connected with or incidental to the foregoing.

2.20 NSSA (ACCIDENT PREVENTION) (WORKERS COMPENSATION SCHEME) NOTICE NO. 68 OF 1990

The National Social Security Authority (NSSA) Act covers occupational health and safety. Workers should work in a healthy and safe environment with protective clothing.

Construction environments have many hazards from which the workers should be protected. The construction workers shall be provided with protective clothing.

2.21   LOCAL AUTHORITY ACT (CAP 265)

Section 160 helps local authorities ensure effective utilization of the sewage systems. Section 170, allows the right of access to private property at all times by local authorities, their officers, and servants for purposes of inspection, maintenance, and alteration or repairs of sewers. The Act under Section 176 gives powers to local authorities to regulate sewage and drainage, fix charges for use of sewers and drains, and require connecting premises to meet the related costs. According to Section 174, any charges so collected shall be deemed to be charges for sanitary services and will be recoverable from the premise owner connected to the facility.

Section 264 also requires that all charges due for sewage, sanitary, and refuse removal shall be recovered jointly and severally from the owner and occupier of the premises in respect of which the services were rendered. This in part allows for application of the "polluter-pays-principle".

2.22   LAND PLANNING ACT (CAP 303)

Section 9 of the subsidiary legislation (Development and Use of Land Regulations, 1961) under this Act requires that before the local authorities submit any plans to then Minister for approval, steps should be taken as may be necessary to give notice to the owners of any land affected by such plans. Public consultation was done in relation to the proposed development at the planning stage

2.23   REGISTRATION OF TITLES ACT (CAP 281)

Section 34 of this Act states that when land is intended to be transferred or any right of way or other easement is intended to be created or transferred, the registered proprietor or, if the proprietor is of unsound mind, the guardian or other person appointed by the court to act on his/her behalf in the matter, shall execute, in original only, a transfer in form F in the First Schedule, which transfer shall, for description of the land intended be dealt with, refer to the grant or certificate of title of the land, or shall give such description as may be sufficient to identify it, and shall contain an accurate statement of the land and easement, or the easement intended to be transferred or created, and a memorandum of all leases, charges, and other encumbrances to which the land may be subject, and of all rights-of-way, easements, and privileges intended to be conveyed.

2.24   LAND TITLES ACT (CAP 282)

The Land Titles Act CAP 282 Section 10 (1) states that there shall be appointed and attached to the Land Registration Court a qualified surveyor who, with such assistants as may be necessary, shall survey land, make a plan or plans thereof, and define and mark the boundaries of any areas therein as, when, and where directed by the Recorder of Titles, either before, during, or after the termination of any question concerning land or any interest connected therewith, and every area so defined and marked shall be further marked with a number of other distinctive symbol to be shown upon the plan or plans for the purposes of complete identification and registration thereof as is herein after prescribed.

2.25   ELECTRIC POWER ACT NO. 11 OF 1997

The Electric Power Act No. 11 enacted in 1997 deals with generation, transmission, distribution, supply, and use of electrical energy as well as the legal basis for establishing the systems associated with these purposes. In this respect, the following environmental issues will be considered before approval is granted:

i. The need to protect and manage the environment and conserve natural resources;
ii. The ability to operate in a manner designated to protect the health and safety of the infrastructure employees and the local and other potentially affected communities.

Under schedule 3 of the Electric Power (licensing) Regulations 2003, it is mandatory to comply with all safety, health, and environmental laws. Moreover, schedule 2 (Regulation 9) of the Electric Power (licensing) Regulations 2003 stipulates that licensing and authorization to generate and transmit electrical power must be supported by the following documents which are approved by EMA.

i. Environmental Impact Assessment Report (EIA)
ii. Environmental Management Plan (EMP)

2.26   BUILDING CODE 2000

Section 194 requires that where sewers exist, the occupants of the nearby premises shall apply to the local authority for a permit to connect to the sewer line and ensure that all wastewater be discharged into sewers.

2.27   PENAL CODE ACT (CAP 63)

Section 191 of the Penal Code states that if any person or institution that voluntarily corrupts or foils water for public springs or reservoirs, rendering it less fit for its ordinary use, is guilty of an offense.

Section 192 of the same Act says it is an offense for any person to vitiate the atmosphere in any place to make it noxious to the health of persons or institutions, dwellings or business premises in the neighborhood or those passing along the public way.

2.28   OCCUPATIONAL HEALTH AND SAFETY ACT (2007)

Before any premises are occupied or used, a certificate of occupation must be obtained from the Chief Inspector. The Act covers provisions for health, safety, and welfare.

2.29   PERSONNEL WELFARE

An adequate supply of both quantity and quality of wholesome drinking water must be provided.

Maintenance of suitable washing facilities and accommodation for clothing not worn during working hours must be provided. Sitting facilities for all female workers whose work is done while standing should be provided to enable them take advantage of any opportunity for resting.

Section 42 stipulates that every premise shall be provided with maintenance and readily accessible means for extinguishing fire, and personnel trained in the correct use of such means shall be present during all working periods.

Section 45 states that regular individual examination or surveys of health conditions of industrial medicine and hygiene must be performed, and the cost will be met by the employer. This will ensure that the examination can take place without any loss of earning for the employees and, if possible, within normal working hours.

Section 55B provides for development and maintenance of an effective program of collection, compilation, and analysis of occupational safety. This will ensure that health statistics, which shall cover injuries and illness – including disabling during working hours – are documented.

2.30   WAY LEAVES ACT (CAP 292)

According to the Way Leaves Act CAP 292 Section 2, private land does not include any land sold or leased under any Act dealing with government lands. Section 3 of the Act states that:

> The government may carry any sewer, drain, or pipeline into, through, over, or under any lands whatsoever but may not, in so doing, interfere with any existing building. Section 8 further states that any person who, without the consent of the Permanent Secretary to the Ministry responsible for works (which consent shall not be unreasonably withheld), causes any building to be newly erected over any sewer, drain, pipeline or other property of the government shall be guilty of an offense and liable to a fine of one hundred and fifty shillings and a further fine of sixty shillings for every day during which the offense is continued after written notice in that behalf from the Permanent Secretary; and the Permanent Secretary may cause any building erected in contravention of this section to be altered, demolished, or otherwise dealt with as he may think fit and may recover any expense incurred by the government in so doing from the offender.

All way leaves shall be granted as and when required by the relevant authorities.

The infrastructure will be governed or will operate within the framework of the above-stated legislations. These will be used in all aspects of the infrastructure from inception to operation.

2.31   DESCRIPTION OF THE ENVIRONMENT SURROUNDING INFRASTRUCTURE LOCATION

2.31.1   INTRODUCTION

This section describes the environment of the infrastructure and where it is situated and planned. The environmental media (climate, soil, water, air, fauna, flora, landscape and interrelationships between the media, social, economic, and cultural heritage) and their description with regard to the EIA are herein described (DEA, 2010).

Both a background as well as baseline data are covered. The biophysical conditions are outlined first as these are the natural resources affected by the infrastructure. This is followed by a section on the land use systems in the area and finally the socioeconomic conditions. It should be noted that while the first chapter provides a concise background of the study area, this section gives a generic description of the Zimbabwe environment.

2.31.2 BIOPHYSICAL PARAMETERS

2.31.2.1 Land level
Land that is level and undeveloped is recommended because it is easier to develop any type of infrastructure upon it.

2.31.2.2 Geology and soils
The geology of the site needs to be known. In the currect infrastructure, the geological structure is comprised of intrusive igneous rocks, largely of gneiss, and granitic in origin. Nearer to the surface this transitions to moderately sandy loams which are whitish-brown in color. In summary the geology of the chosen area is conducive to building and road construction, owing to its underlying igneous and the granite geological structure. The soils are generally well-drained.

2.31.2.3 Drainage
The site generally drains to the east toward a railway line. The drainage system comprises a level area.

2.31.2.4 Climate
The area experiences some rainfall in summer. The rainfall is associated with thunder and heavy downpour which lasts about 5 hours. It rarely rains, but when it does, it's heavy and unpredictable.

2.31.2.5 Vegetation
Vegetation is dominated by scattered Mopani trees, baobab trees, thorn trees, and low shrubs of a secondary nature for parts of the area under cultivation, and transitions to fairly dense forest. There will be need for the beneficiaries of stands to plant trees, grass, and lawn in those areas currently under cultivation. No protected tree species were found in the area.

2.31.2.6 Wildlife
The infrastructure area is covered by trees which provide habitat for bird populations in the area. There are no endemic mammals in the planning area. Small animals and reptiles were seen in the area, indicating a hub for a variety of bird and animal species.

2.32 LAND USE SYSTEMS

The proposed infrastructure site is mostly open state land.

2.33 TELEPHONE LINES

Telephone lines run parallel to the existing highway and can be extended to reach the residential and commercial stands.

2.34 WATER AND SEWER

Boreholes (wells) are recommended to be the source of water.

The biophysical conditions and the land systems in the area indicate that the site is suitable for the proposed development. Spaced dryland grass covers the flat infrastructure area.

2.35 SIGNIFICANCE OF ENVIRONMENTAL IMPACT ANALYSIS AND MITIGATION MEASURES

2.35.1 INTRODUCTION

The purpose and procedures of carrying out the EIA before the infrastructure starts are to ensure that the development options under consideration are environmentally sound and that any environmental consequences are determined early and taken into account in the infrastructure design. The assessment will be separated into the biophysical and socioeconomic components for the ease of analysis. However, various issues have aspects relating to both components, so, consequently, some overlap exists. Several impacts, both positive and negative, will be experienced during the planning, construction, and operation phases of the infrastructure. The impacts arise out of the envisaged process as set out in subsequent tables in each phase. The potential impacts that represent damage or loss of a resource are presented in this chapter, whilst administrative or social implications are presented in the following chapter on socioeconomic impacts.

Environmental impacts which will be encountered during the planning, construction, and operational phases will be discussed.

The significance of ecological impacts will be evaluated primarily based on the following criteria (Tiwari et al., 2014):

   i. Habitat quality
  ii. Species affected
 iii. Size/abundance of habitats/organisms affected
 iv. Duration of impacts
  v. Impact significance
 vi. Reversibility of impacts
vii. Magnitude of environmental changes

2.35.2 PLANNING PHASE

This concerns the current state of the environment in the infrastructure area and refers to the set of biophysical parameters, which form the baseline data as described previously. Most of their benefits and costs to the environmental, socioeconomic, and cultural fabric of the proposed development area and its neighborhood will not continue into the operation and maintenance phases. They are also confined to the specific areas, namely, areas of pegging, clearance, and site visits as well as related activities. Some of the activities involved in the planning phase include the following:

i. Consultation of relevant stakeholders
ii. Environmental impact assessment by the consultant
iii. Detailed survey by the consultant

The only activity that has any significant environmental impact in the planning phase is the detailed site survey, which involves limited clearing of vegetation.

The survey team will temporarily affect the study area by going into the area, pegging, and opening lines. Temporary shelter will be put in one place for the survey team, but this will not last long.

2.35.2.1 Impact on Vegetation

The removal of vegetation cover is perhaps the largest single impact that will occur in the planning phase. The stripping of the vegetation cover will be a result of four principal activities:

i. Site pegging
ii. Survey of the site
iii. Site inspection by engineers
iv. The initial clearing work is to be carried out in and around the stand (McTavish, 2001).

The major causes of these impacts will be related to the movements, pegging, servicing, and development of access roads. Associated with the removal of vegetation cover are the exposure of the top soil and the possibility of soil erosion. This is particularly serious on sloppy ground, some of which has above average erosion hazards. The geographic extent of this is, however, limited. This will be confined only to limited areas where the soils are fragile. However, heavy convectional summer rainfall is likely to wreak havoc on any exposed soils.

Mitigation

Minimizing vegetation clearance to when it is necessary is the soundest mitigation measure possible. We also recommend the conferment of clearance to the absolutely necessary during the preliminary visits. Over clearance is discouraged. Runner grass must be planted on marginal areas.

Riverine vegetation should be left undisturbed, as this will reduce the effect of erosion.

2.35.2.2 Hydrological Impacts
Mitigation

Minimization of the vegetation clearance and soil compaction and the planting of alternative tree and grass species along the stream and drainage channels leading to the stream are strongly recommended. There is a need to raise the levees of the waterway especially the eastern side, to avert flooding the eastern residential stands situated along the waterway.

Impacts to Wildlife

Bird species of conservation importance were found in the area.

The survey revealed that the bird population in the study area is fairly high. The removal of vegetation, though minimal in the preliminary stage, must be contained. Noise from the movement of vehicles and human sound is likely to disturb the movement and feeding pattern of the birds and other smaller organisms which are also significant in numbers. The possibility of some animals protecting their territories resulting in injuries can also not be ruled out (e.g. snakes).

Mitigation

Disturbance to the ecosystem should be kept to a minimum. Avoid oil, waste, and other forms of pollutant leakages into the area. Vigilant precaution should be taken to protect all animals and their young ones living in the ecosystem.

Socioeconomic and Cultural Impacts

Socioeconomic changes in the areas of health, transport, and other businesses are predicted. These will include the employment of locals as temporary casual laborers assisting the surveying team (REMA, 2016) and commerce between the team and local shops. Despite this, resentment by some residents to the infrastructure cannot be ruled out, especially given that the proposed fields are those which locals might have been cultivating illegally. The socioeconomic activities impact analysis is represented in Table 2.2. The aesthetic impact analysis is represented in Table 2.3.

2.36   SOCIAL IMPACT ASSESSMENT

2.36.1   INTRODUCTION

Social impacts assessments will be done to establish the perception of interested and affected parties in the infrastructure development. The interests and objection of

TABLE 2.2
Socioeconomic and Cultural Impact Analysis

Effect Magnitude	Duration	Probability	Intensity
Large	Short/Periodic	Certain	Moderate
Large	Long/Continuous	Probable	High
Large	Short/Periodic	Probable	Moderate
Small	Long/Continuous	Certain	Moderate
Small	Short/Periodic	Certain	Low
Small	Long/Continuous	Probable	Low
Small	Short/Periodic	Probable	Low

TABLE 2.3
Aesthetic Impact Analysis

Category	Impact	Cause	Magnitude of Impact	Duration and Probability of Occurrence	Significance Rating
Aesthetic impacts	Construction of houses Access roads	Structures spread across the land. Access roads crisscrossing the terrain	Small magnitude: Aesthetic impact constrained to building areas only	Continuous Duration: structures will be permanently there as long as the infrastructure remains functional. Probability: Certain to happen	Moderate

the neighboring communities will be evaluated accordingly. This is primarily done through questionnaires specially designed to collect relevant information.

2.36.2   OBJECTIVES OF PUBLIC CONSULTATION

The objectives of the public consultation are to: establish all the relevant aspects of the infrastructure and community knowledge, perceptions and expectations; ensure that every stakeholder is aware of the development; avoid potential land conflicts; and consider all value-adding contributions.

2.36.3   METHODOLOGY

A questionnaire survey was used to collect information from the interested and affected parties. The questionnaire was designed specifically for the typical infra-structure and is user friendly. Copies of the consultations which were done are attached to the report.

2.36.4   PUBLIC INVOLVEMENT OF INTERESTED AND AFFECTED PARTIES

Five randomly chosen households were consulted by field workers. Their perceptions were documented and are part of this report.

2.36.5   SOCIAL IMPACT ASSESSMENT

The infrastructure has several potential positive and negative impacts from a social perspective. The way of life will change due to increase of population in the area. Problems such as prostitution and theft may arise. Health risk may also become part of the community problems. The current infrastructure and utilities may fail due to overtaxing by population and lead to disease outbreak if uncontained. From the vegetation perspective, there will loss of biodiversity. However, among positive impacts is a decrease in household backlog and organized housing facilities.

2.36.6   MITIGATION IN CONSTRUCTION PHASE

Mitigation measures in the construction phase for the physical/biological environment are indicated in Table 2.4.

Socioeconomic mitigation measures during the construction phase are indicated in Table 2.5.

2.36.7   MITIGATION IN OPERATION AND MAINTENANCE PHASE

Mitigation measures during operation and maintenance for the physical/biological environment are indicated in Table 2.6.

Mitigation measures during operation and maintenance in the socioeconomic environment are indicated in Table 2.7.

TABLE 2.4
Physical/Biological Mitigations at Construction Phase

Activity/Action	Mitigation measures
Deforestation	Remove as little vegetation as possible and localize to construction areas
Erosion	Remove as little vegetation as possible and localize to construction areas
Sedimentation in water bodies	Ensure proper drainage during construction
Loss of biodiversity	Vegetate exposed surfaces
Endangered species decimation	Relocate endangered species to alternative site or construct on another land with less negative impact
Land use conflicts	Meet with community leaders to manage land use conflict
Waste management problems	Ensure proper waste management and availability of landfill site

TABLE 2.5
Socioeconomic Analysis and Mitigations at Construction Phase

Activity/Action	Mitigation measures
Heavy vehicle traffic and congestion along existing roads	Direct heavy traffic around already built-up areas
Noise and disturbance from construction	Avoid routes passing through residential sites
Increased incidence of accidents, disability	Educate drivers, enforce speed limits
Increase in communicable disease from workers and surrounding communities, changes in mortality, increased health costs	Implement safety, health, and awareness programs for employees
Community breakdown	Civic education
Increase in number of people employed in the area	Employ local people
Destruction of sites of historical or cultural importance	Avoid routes passing through sites of cultural importance

2.37 ENVIRONMENTAL MANAGEMENT PLAN, IMPLEMENTATION, AND TRAINING

In a new area of housing infrastructure, the physical environment has its own particular and specific attributes. The area is adjacent to a highway and bordered by a railway line to the eastern side. The legal distance between the mentioned aspects were observed prior to allocation of the site. It is therefore acknowledged that a degree of interaction will be experienced and the appropriate mitigatory measures should be ensured and implemented.

It should be acknowledged that the public has to be well trained and informed of the potential environmental problems they might face as a result of their daily activities with regard to the physical features of their area and the surrounding area. Waste

TABLE 2.6
Physical and Biological Analysis and Mitigation at Operation Stage

Activity/Action	Mitigation measures
Pollution of water and soil from incorrect disposal, dumping	Ensure water and soil pollution program is followed and waste disposal is done appropriately
Breakdown in drainage system	Consider the utilization of poly vinly chloride pipes
Urban agriculture	Adhering to planning, design, and zoning standards.
Change in hydrology	Ensure proper drainage during construction
Changes in microclimate	Ensure no burning of waste
Air pollution	Ensure no burning of waste
Solid waste disposal	Local authority to ensure proper functioning of domestic solid waste removal system

TABLE 2.7
Socioeconomic Analysis and Mitigations at Operation Stage

Activity/Action	Mitigation measures
Traffic issues	Educate drivers, enforce speed limits
Pressure on local infrastructure	Ensure adequacy of public utilities (water, storm water, sewage, electricity, phone lines) in relation to development
Health issues	Ensure access to emergency medical care in case of serious injury to workers or community
Effect of area aesthetics	Vary housing designs
Risks/hazard from pollution, accidents	Educate public and ensure availability of emergency medical care

handling should be appropriately administered and the basic foundation adhered to. Separation at source would be implemented.

2.37.1 CONSTRUCTION PHASE

The construction company engaged to build the infrastructure must respect the rules of waste management and ensure a safe and clean working environment which does not leave foot prints.

2.37.2 OPERATION PHASE

The operation of the facility includes but is not limited to:

i. Water use and its system
ii. Sewerage use and its systems
iii. Waste management system

TABLE 2.8

Environmental Management and Monitoring Plan for Housing Construction

Environmental Management and Monitoring Plan

Impact for Mitigation	Management Action	Responsibility	Monitoring and Auditing	Time Frame
Housing area generated waste	**Target: Minimize waste impacts** **Planning Phase** Dust suppression measures to be included in operating plan **Operating and Remedial Measures** Gravel roads to be wetted when dusty. Complaints register to be kept. Establish a waste collection monitoring committee. Road signs on the road. Maintenance of traffic warning signs in position, particularly at the intersection with the main road. Speed limit warnings before the turn offs. Maintenance of access road and on-site road to be in trafficable conditions at all times, including wet weather. **Standards** Minimum requirements for waste management	Consultant City Council City Council City Council Ministry of Transport City Council Site Operator	Responsibility City Council City Council To inspect the condition of the roads and waste management system. **Monitoring Committee** To assess the complaints register	Through the period of planning During the operation period Daily basis Daily basis Once every five years Weekly based Daily
Sewer system management and operation	**Target: Minimize nuisances** **Planning Phase** Measure to control nuisance to be included in the operating plan (odors and blocked drainage) **Operating and Remedial Measures** Ensure proper sanitary practice of maintenance Ensure no sewer overflow due to blocking. Complaints register to be kept	Site Operator	Responsibility City Council **Monitoring Action** Inspections must be held annually, at a minimum.	Throughout the year. Daily basis Daily basis Annually

(Continued)

TABLE 2.8 (CONTINUED)
Environmental Management and Monitoring Plan for Housing Construction

Environmental Management and Monitoring Plan

Impact for Mitigation	Management Action	Responsibility	Monitoring and Auditing	Time Frame
Record keeping	Target: Prevent pollution of water resources **Planning Phase** No measures **Operating and Remedial Measures** Keep records of all waste generated Categorize waste by the number of loads, defined by mass and type Record keeping on both daily and a cumulative basis Site reporting structure and lines of authority to be clearly defined	Site Operator Site Operator	Responsibility City Council Monitoring Committee **Monitoring Action** City Council and Waste Monitoring Committee to regularly inspect waste records	Daily basis
Water quality Monitoring and site drainage	**Target: Prevent pollution of water resources** **Planning Phase** Indicate drainage or runoff in design Indicate drainage for contaminated water in design Indicate drainage for clean water in design	Consultant	Responsibility City Council **Monitoring Action**	On agreed routine. Maybe weekly depending on City Council routine
Security	**Target: Control criminal activities** To minimize the risk to the public **Planning Measures** Indicate fencing and firebreak requirements **Operating and Remedial Measures** Maintain the security boundaries	Consultant	Responsibility City Council Security monitoring committee	Daily

(Continued)

TABLE 2.8 (CONTINUED)
Environmental Management and Monitoring Plan for Housing Construction

	Environmental Management and Monitoring Plan			
Impact for Mitigation	**Management Action**	**Responsibility**	**Monitoring and Auditing**	**Time Frame**
Continuous disposal of waste in the environment	**Target: Prevent further degradation of the environment**	Consultant	**Responsibility**	
	Planning Phase	Site Operator	City Council	Daily
	Indicate rehabilitation Plans. Rehabilitation Plan to indicate that only indigenous grasses to be planted in areas cleared.	City Council	**Monitoring Action**	Daily
	Operation and Rehabilitation		Inspection of vegetation	
	Hydro seed areas with indigenous grass species after clearing waste or landscaping.			
Waste collection auditing	**Target: To ensure maintenance of acceptable standards**	Consultant	**Responsibility**	
	Planning Phase	Site Operator	City Council	Daily
	Indicate frequency and type of audits to be carried out	City Council		Daily
	Operating Measures			
	Audit residential location on regular basis			
Houses standards	**Target: To ensure maintenance of houses to acceptable standards**	Consultant	**Responsibility**	At construction and
	Planning Phase	Site Operator	City Council	when new
	List of facilities to be maintained (road, drains, contaminated water pond, house, and fencing)	City Council	**Monitoring Action**	developments are
	Operating Measures		Inspect condition of facilities and	implemented
	Maintain all facilities including the access roads		houses	

(Continued)

TABLE 2.8 (CONTINUED)
Environmental Management and Monitoring Plan for Housing Construction

Environmental Management and Monitoring Plan

Impact for Mitigation	Management Action	Responsibility	Monitoring and Auditing	Time Frame
Health and safety risks	**Target: To minimize the risk of illness or diseases (cholera and typhoid) to residents**	Consultant	**Responsibility** City Council NSSA	Daily
	Abiding by NOSA or OSHA requirements	Site Operator		
	All health and safety measures as stipulated in current legislation to be followed at all times.	Site Operator	**Monitoring Action** Inspection of health and safety standards on site	
	Planning Phase			
	Stipulation of health and safety requirements	Site Operator		
	Operating Measures	City Council		
	Appropriate safety awareness signs to be prominently displayed in the area	Site operator		
	Fire extinguishers to be provided where necessary			
	Preparation of an emergency plan for any unforeseeable emergencies such as accidents/outbreak of infectious diseases			
	Provision of clean portable water for use by residents			
Conflict management and maintenance personnel	**Target: To minimize public conflicts regarding residential area and free space**	Consultant	**Responsibility** City Council Housing committee	Daily and when matters arise
	Demarcate yards.	Site Operator		
	Planning Phase	Site operator		
	Specify ownership strategy	City Council	**Monitoring Action** Assessment of who uses and owns which area	
	Specify measures to minimize conflict between the communities and the municipality	City Council		
	Operating Measures			
	Employment of locals for unskilled and semiskilled positions			
	Provision equitable employment opportunities for local community			
	Hiring of local contractors to work at the area			
	Establish a housing committee			
	Liaison between the councilors and the Council			

Furthermore, it is recommended that the national legal framework according to -Table 2.9 is implemented (World Business Council for Sustainable Development, 2005).

TABLE 2.9
Implementation of the Legislation Framework

Instrument Legal Framework	Aspect for Compliance	How to Comply
Environmental Management Act Ch. 20:23 Sections 97 to 108 of Part XI	Infrastructures which require an EIA process	All infrastructures to be done in the area which require an EIA should be done prior to implementation
Environmental Management Act (20: 27) of 2003	Provision of waste management service	It is the City Council's mandate to provide waste management system and waste disposal system
National Parks and Wildlife Act Section 24	Provides for the protection of plants, animals, and any wildlife that may be in danger from hunting or gathering	Ensure that programs are there to address the protection of plants and animals. Responsible committee should implement them
Road Traffic Act Section 38	Building construction should not encroach upon a restricted road	Construction standards should be observed
Regional, Town and Country Planning Act Chapter 29:12	Construction works	Construction works should be according to plans and directed by competent person
Public Health Act	The Act prohibits creation of nuisance	City Council to provide sanitary convenience: waste, water, and sewerage system
The Explosives Act (10.08)	Blasting	Developer to ensure that correct procedures are done before, during, and after blasting activities
Forestry Act Sections 38 and 39: Preservation of Trees of Forest Procedure	Tree-cutting	Enforce respect of nature by prohibiting tree felling
Pneumoconiosis Act Cap 15:09	Protection of site workers	Site workers shall be provided by developer with protective clothing to avoid inhaling dust
The Atmospheric Pollution Act (1971)	Prevention and control of air pollution by noxious or offensive gases, dust, smoke, and internal combustion fumes	The developer shall dampen the ground at implementation stage to avoid dust. Construction workers will wear protective clothing to avoid infection
Mines and Mineral Act 1976 CAP 21:05	Location of the residential area in relation to the mine claims	Ensure a distance of 450 mm is observed between the mine claims and residential area

(Continued)

TABLE 2.9 (CONTINUED)
Implementation of the Legislation Framework

Instrument Legal Framework	Aspect for Compliance	How to Comply
The Hazardous Substances and Articles Control Act (1971)	Dumping of industrial poisons, pesticides, and other dangerous substances and waste, including radioactive materials	The developer and the City Council will ensure that no hazardous substances are dumped in the infrastructure area or at a site not suitable for hazardous waste
Water Act 20:24 (1998) Sections 67-71	Ensuring that water resources management is consistent with broader National Plan	Adhere to acceptable standards by water analysis
Water Act 20:24 (1998) CAP 20:27 SI No 2007	The maximum permissible concentration of chemical constituents permissible in water which is discharged or disposed of in a Zone 1 or Zone 2 catchment shall be observed	Adhere to acceptable standards by water analysis
Labor Act, Chapter 28.01	Fundamental rights of employees	The City Council and developer to observe the labor act by ensuring workers' rights
NSSA (Accident Prevention) (Workers Compensation Scheme) Notice No. 68 of 1990.	Workers to work in a healthy and safe environment with protective clothing	The construction workers shall be provided with protective clothing by the developer and the City Council
Local Authority Act (CAP. 265) Section 160 Section 170 Section 176 Section 174 Section 264	Effective utilization of the sewages systems Right to access to private property at all times by Local Authorities, its officers, and servants for purposes of inspection, maintenance, and alteration or repairs of sewers Empowers Local Authority to regulate sewage and drainage, fix charges for use of sewers and drains, and require connecting premises owner to meet the related costs Any charges so collected shall be deemed to be charges for sanitary services and will be recoverable from the premises owner connected to the facility.	Local authority to observe its responsibilities and mandate. Tenants to own their obligations

(Continued)

TABLE 2.9 (CONTINUED)
Implementation of the Legislation Framework

Instrument Legal Framework	Aspect for Compliance	How to Comply
	Also requires that all charges due for sewage sanitary and refuse removal shall be recovered jointly and severally from the owner and occupier of the premises in respect of which the services were rendered. This in part allows for application of the "polluter-pays-principle"	
Land Planning Act (CAP. 303) Section 9 of the subsidiary legislation (Development and Use of Land Regulations, 1961)	Public Consultation	Before the local authorities submit any plans to the Minister for approval, steps should be taken as may be necessary to alert the owners of any land affected by such plans. Public consultation was done in relation to the proposed development at the planning stage
Building Code 2000 Section 194	Permit to connect to sewer line	Where sewers exist, the occupants of the nearby premises shall apply to the Local Authority for a permit to connect to the sewer line and all the wastewater must be discharged into sewers
Occupational Health and Safety Act (2007) Before any premises are occupied or used, a certificate of occupation must be obtained from the Chief Inspector. The occupier must keep a general register	Occupation of premises	Before any premises are occupied or used, a certificate of occupation must be obtained from the chief inspector. The occupier must keep a general register
Way Leaves Act CAP 292 Section 2	Private land	Legislation in this regard should be followed.

These aspects should be ensured that they are in working order at all times, and the public should be informed of the current and future developments. Expansion of the area and immigration should be met with expansion of such services so that health hazards are not generated.

2.38   ENVIRONMENTAL MANAGEMENT AND MONITORING PLAN

The Environmental Management and Monitoring Plan for the housing infrastructure is indicated in Table 2.8.

Furthermore, it is recommended that the national legal framework according to Table 2.9 is implemented (World Business Council for Sustainable Development, 2005).

2.39   CONCLUSIONS AND RECOMMENDATIONS

2.39.1   CONCLUSION

The establishment of green buildings is important for environmental protection. The approach to the provision of housing service must always consider the environmental impacts and management plans.

2.39.2   RECOMMENDATIONS

All new housing establishments must have integrated solid waste management systems. These schemes must also consider renewable energy sources, like biogas (from waste) and solar power. Adoption of green buildings will promote sustainability.

REFERENCES

DEA (2010). Companion to the National Environmental Management Act (NEMA) Environmental Impact Assessment (EIA) Regulations of 2010, Integrated Environmental.

EPA (2008). Environmental Protection Authority, Environmental Guidance for Planning and Development (May 2008). *Guidance Statement No. 33.*

Management Guideline Series 5 (2010). Department of Environmental Affairs (DEA), Pretoria.

McTavish, J. J. (2001). A Review of Jamaican and International Noise Standards. For Natural Resources Conservation Authority (2001).

REMA (2005). Sector Guidelines for Environmental Impact Assessment (EIA) for Housing Projects in Rwanda.

REMA (2016). Environmental and Social Impact Assessment (ESIA) Report. Development of Urban Infrastructure in six Secondary Cities of Rubavu, Rusizi, Musanze, Muhanga, Huye and Nyagatare of Rwanda, and the City of Kigali, January, 2016.

Tiwari, Vivek Kumar, Dutta, Venkatesh and Yunus, M. (2014). A Comparative Study of Environmental Impact Assessment Reports of Housing Projects of Lucknow City, Uttar Pradesh, India. *G-Journal of Environmental Science and Technology*, ISSN (Online), 2322–2328.

United Nations (2000). The World Commission on Environment and Development.

World Business Council for Sustainable Development (2005). Environmental and Social Impact Assessment (ESIA) Guidelines, 54.

3 Environmental Impact Assessment for Coal Bed Methane Drilling

3.1 INTRODUCTION

Zimbabwe is endowed with a vast range of mineral resources which include precious and semi-precious minerals as well as various fuel deposits. The latest statistics show that Zimbabwe is the richest country in the world in mineral quantity per capita, making mining and minerals processing critical to improving the country's economy and raising its quality of life standards. The 2012 Mid-Term Fiscal Policy review states that the mining sector contributes close to 11% to Zimbabwe's gross domestic product (GDP), over 50% of total earnings, and approximately 45 000 jobs. Despite this enormous wealth potential, the lack of investment in electric power generation combined with high consumer demand has crippled the country's ability to fully exploit these natural resources. The electricity crisis is a constant worry for leaders in sub-Saharan Africa and the Zimbabwe Energy Regulatory Authority (ZERA), as the country is rocked by continual power outages.

Natural gas has the potential to address Zimbabwe's power crisis by providing heat as well as lighting, and the country is exploring the possibility of tapping into coal-bed methane (CBM) gas abundant in parts of Matabeleland North Province. According to reports, it is estimated that the methane gas in Matabeleland is 95% pure and lies between Hwange and neighboring Botswana (McCauley, 2000). However, the extractable quantity still needs to be determined.

The Zimbabwean economy is dominated by the mining sector. Although multinational companies predominate, there are some small to medium and artisanal miners. Mining and associated activities result in income multipliers arising from both direct and indirect employment. At the national level, the mining sector has a direct positive impact through payment of tax revenue. Indirect positive impact arises from income taxes on employment, personal income, profits of local business, major suppliers, and purchase of goods and services. Very few of the local miners have ventured into coal bed gas extraction due to the high cost.

Exploitation of coal bed gases is associated with limited negative impacts. As an alternative to hydro power generation, coal bed gas is cheaper to mine and is environmentally friendly (APSA, 2001). The gas can be harnessed and used in homes, restaurants, and schools, among other facilities which rely heavily on electricity for heating and lighting. In this regard, the company interested in CBM exploration has a high potential of improving Zimbabwe's power grid capacity and the country's sustainability.

3.2 TERMS OF REFERENCE

In January 2007, Zimbabwe's Environmental Management Agency (EMA) amended the Environmental Management Act CAP 20:27 of 2002 to minimize environmental degradation caused by illegal mining activities. EMA regulations state that an EIA should be conducted, approved, and certificate issued prior to implementation of the development projects including mining and associated activities. Accordingly, in order to meet the requirements of the Environmental Management Act CAP 20:27 of 2002; Section 97, the proponent will contract an environmental consultancy to prepare an EIA report. The EIA outlines environmental management practices that ensure that adequate mitigation or protection measures are incorporated into the project design, implementation, and decommissioning phases in order to allow for sustainable coal gas extraction activities. The primary data from the report is based on ground truthing, site scoping, environmental characterization, and desktop research. Secondary data for the report is provided by proponent by way of disclosure of potential, production-quantified performance data, supply analysis records, and mine owner's history report. In line with the terms of reference (TOR), the prospecting EIA report will include:

i. A description of the entire project, including associated infrastructure requirements
ii. An outline of the various approvals required for the project to proceed
iii. Descriptions of the existing environment, particularly where this is relevant to the assessment of impacts
iv. Measures for avoiding, minimizing, managing, and monitoring adverse impacts, including a statement of commitment to implement the measures
v. Rigorous assessment of the risks of adverse and beneficial environmental impacts arising from the project and relevant alternatives on environmental, social, and economic values relative to the scenario
vi. Any information derived from baseline and predictive studies, the required extent of which will be commensurate to risks
vii. A description of stakeholder consultation undertaken
viii. Responses to issues which arose during public and stakeholder consultation

3.3 ENVIRONMENTAL LEGISLATIVE REQUIREMENTS

The Zimbabwe government, as stipulated in the Environmental Management Act Cap 20:27 of 2003, made it mandatory for EIAs to be undertaken for all major development projects that are likely to have a negative impact on the environment. CBM gas prospecting, extraction, and associated activities, as specified in Section 97 of the Act, require Environmental Impact Assessment and Environmental Management Plans (EMPs). The Environmental Management Agency (EMA) is responsible for monitoring projects and advising on environmental concerns. Proponent commits to complying with laws that provide for the protection of Zimbabwe's environment, which includes:

3.3.1 ENVIRONMENTAL MANAGEMENT ACT (2003) CAP 20:27

Under this Act, the following sections are relevant for the current study:

Section 97 lists projects that require EIA/EMP undertakings; and this project falls in the category of projects that require an EIA/EMP.

Under Section 98, the developer is required to submit a prospectus to the Director General containing information regarding the EIA and state whether the project is the prescribed activity

Under Section 100, the report will be reviewed by the Director General within 60 days from the date of submission and, if the report is approved, a certificate valid for 2 years will be issued and can be renewed.

3.3.2 PUBLIC HEALTH ACT 1971 CAP 15:09

The Act provides for:

i. Supply of suitable water
ii. Prevention of pollution of water resources
iii. Sanitation and control of infectious diseases

The company interested in CBM exploration will provide workers with personal protective clothes and ensure the working area is safe for workers by providing appropriate procedures (IUCN, 2009). Monitoring and supervision will be done to ensure worker safety.

3.3.3 PNEUMOCONIOSIS ACT CAP 15:09

The Act provides for the control and administration of persons in dust occupations. A medical bureau is established in terms of this Act to *inter alia* perform medical assessment and record incidences of pneumoconiosis. Workers are issued certificates that certify fitness to work in dust conditions.

3.3.4 THE ATMOSPHERIC POLLUTION ACT (1971)

Under the Act, uncontrolled production and disposal of artificial waste are prohibited. The Act provides for the prevention and control of air pollution by noxious or offensive gases, dust, smoke, and internal combustion fumes. While it is not expected that high quantities of gases will be emitted, the company interested in CBM exploration will take measures to ensure the safety of the workers and the environment by making sure no gases will be emitted into the environment. A gas detector will be used to determine gas emissions.

3.3.5 MINES AND MINERAL ACT 1976 CAP 21:05

The Act calls for mandatory consultation with the occupier or owner of the land on which the project is located. The Mines and Mineral Amendment Bill 2004, Section

157 proposes that miners, as far as reasonably practical, rehabilitate the environment affected by their operations to its previous natural state. Locals and interested or affected parties were consulted in regard to the development.

3.3.6 Forest Act CAP 19:05

The Minister has wide powers that relate to restrictions on import of specified trees, the suppression of tree diseases, and noxious and non-indigenous trees. The rehabilitation of the areas will be done by use of indigenous trees and grass.

3.3.7 The Hazardous Substances and Articles Control Act (1971)

Regulates the dumping of industrial poisons, pesticides, and other dangerous substances and waste including radioactive material. The company interested in CBM exploration will collect all waste and dispose of them in designated areas.

3.3.8 Water Act 20:24 (1998)

Part IV of the water act makes provision for the control of water pollution and protection of water resources. In Sections 67–71 of the Act, provision is made insuring that water resources management is consistent with the broader National Plan CAP 20:27 SI No 2007. The company interested in CBM exploration will ensure that no streams or rivers will be polluted due to its activities. Point drilling uses water but does not pollute it; the water will still be environmentally sound after use.

3.4 OVERVIEW OF PROJECT

The drilling project, which is located in Copper Queen, focuses on determination of the occurrence and quality of the Permian age Ecca Coal gas (methane) measures of the mid Zambezi Basin. Copper Queen is located in Gokwe, the largest district in Zimbabwe, which lies to the northwest of the country. It is in the country's agro-ecological Region Three, receives 819 mm rainfall annually, and experiences an annual average temperature of 26°C. Cotton is the major agricultural income earner for farmers in the Gokwe District. The district has experienced an unprecedented cotton boom since independence in 1980. The district's contribution to national cotton output fluctuated between 31 and 51% of that in other communal areas. The number of registered cotton growers in Gokwe increased from 24 800 in 1980 to 55 700 in 1984–1985. Cotton production grew faster than all other crops because of the suitability of farming conditions.

Various types of coal contain different amounts of methane gas due to their properties, such as porosity. The porosity of coal bed reservoirs is usually very small, varying from 0.1% to 10%. Coal bed gas adsorbs to the coal bed. Adsorption capacity of coal is defined as the volume of gas adsorbed per unit mass of coal usually expressed in standard cubic feet (SCF), (the volume at standard pressure and temperature conditions) gas/ton of coal (Swan et al., 1994). The capacity to adsorb depends on the rank and quality of coal. The range is usually 100–800 SCF/ton for most coal seams. Most of the gas in coal beds is in the adsorbed form. When the reservoir is put into production, water in the fracture spaces is pumped off first to enhance the desorption of gas from the matrix.

This report focuses mainly on the prospecting activities. To extract the gas, a steel-encased hole is drilled into the coal seam (100–1500m below ground). As the pressure within the coal seam declines due to natural production or the pumping of water from the coal bed, both gas and "produced water" come to the surface through tubing (Wijffels et al., 1996). Then the gas is sent to a compressor station and into natural gas pipelines. The produced water is either re-injected into isolated formations, released into streams, used for irrigation, or sent to evaporation ponds. The water typically contains dissolved solids such as sodium bicarbonate and chloride but varies depending on the formation geology.

3.5   PROPOSED DRILLING PROGRAM

The proposed wells will be drilled after acquiring the EIA certificate. An approach of vertical exploration will be implemented. The day-to-day management of the drilling operations will fall under the total responsibility of the company interested in CBM exploration. All decisions regarding the drilling operation and execution of this program will be its responsibility based upon the input from both the respective company drilling supervisors/drillers and the on-site geologist. The potential drilling program is detailed below:

   i. Mobilization approximately which may include: 15 Rig, personnel and camp to site
  ii. Check and confirm well coordinates
 iii. Setup equipment ready for drilling to commence
 iv. Conduct pre-spud meeting, cover site-specific risks, and confirm drilling plan. Reemphasize the environmental sensitivity and ensure compliance is maintained against approvals
  v. Drill 12" hole with air/mist to a minimum ~12 m (decision based on drilling supervisor permission, availability of water, and local ground conditions)
 vi. Install 10" OD NB casing and cement in place. Wait on cement to set and harden (minimum 6 hours) to ensure a good cement bond before proceeding
 vii. Hammer drill 10" hole to approximately 2–3 m into competent basalt (as determined by well-site geologist and drilling supervisor)
viii. Install 8" OD NB welded steel bevelled casing
 ix. Install cement head and cement to surface. Cement with minimum 50% excess cement on annular fill requirement (excess to be increased if hole conditions are poor). Cement slurry is to be mixed at a ratio of one 40kg bag of GP and/or builders cement to 25 liters of water and have a measured S.G. of no less than 1.65. Perform annular surface top up if required
  x. Displace cement with water and shut in
 xi. Remove cement head and install wellhead flange and diverter/flow-line
 xii. Hammer drill 6.5" hole through the basalt and into the underlying aquifer sandstone section of the Ntane Sandstone

xiii. Review hole conditions and water flow. Barring significant hole conditions that would prevent drilling to continue, pull out of hole to replace 6.5" hammer with 6-1/8" tricone bit. Run in hole and commence mixing bentonite mud system. Circulate hole to mud and condition mud to correct properties. Once hole is stabilized and mud system optimized.

xiv. Contingent operation: Should hole conditions indicate an unstable hole at the base of the basalt and top Ntane Sandstone, pull out of hole with 6.5" hammer and pick up 8" hammer to open hole up. Open hole to TD 8". Ensure hole is clean and pull out of hole. Rig down diverter and run 6.5" OD NB welded casing to TD. Top set at least 10 sacks of cement to seal the annulus between the 8" and 6.5" casing string. Weld centralizer gussets to maintain and stabilize the 6.6" casing inside the 8" casing. Refit the 8" wellhead flange and diverter, flow-line.

xv. Make up 6-1/8" tricone bit and run in hole. Drill 6-1/8" hole TD and circulate and condition hole for logging

xvi. Run wireline logs as per program

xvii. Plug and abandon, run in hole with open-ended pipe to set abandonment plugs. Set 50 m plug across the top Ecca Coal Formation top (25 m below and 25 m above). Set second 50 m plug across the casing shoe and 10 sack surface plug

3.6   OPERATIONAL WASTE

Normal drilling operations generate the following types of waste (Renewable Natural Gas, 2004):

   i. Drill cuttings, discharged continuously during drilling
   ii. Water-based drilling fluids/muds, generally discharged at completion of the well
   iii. Sewage, gray water, and putrescible wastes discharged at site
   iv. Cooling waters, discharged continuously during drilling
   v. Domestic and industrial solid wastes and hazardous solid and liquid wastes, collected and segregated for transport to appropriate disposal site at intervals during drilling
   vi. Engine waste and gas, which will be collected and transported to appropriate disposal

The drill rigs will have containment zones and bunding in all areas where gas products are stored and gas residues will be stored in drums for disposal at authorized sites. Minor oil spills will be washed with biodegradable detergents and polluted water will be collected in a settling tank for future disposal at authorized sites.

3.7   WELL CONTROL PROCEDURES

The well control procedures are based on three key elements:

   i. Thorough assessment of the geology and formation pressures prevalent in the area

ii. Design of the drilling fluid program

iii. Well control procedures used by the drilling contractor

The drilling program will fully incorporate these three key well control elements to provide an industry 'best practice' approach to well control. This will include training and accreditation to both the drilling contractor and personnel of the company interested in CBM site exploration.

3.8 DRILLING SAFETY

The positioning and operation of the drill rigs will be closely supervised by the company drill supervisor. During the drilling program, a temporary safety exclusion zone with a radius of 500 m around the drill rigs will be declared and appropriately gazetted. Wells will be designed and engineered to approved standards to ensure that well pressures remain within the safety limits. Blow-out preventers (BOPs) will be used to contain pressures in excess of those encountered in earlier wells.

Casing sizes and lengths and the intervals where the hole is cement-sealed around the casing will be selected to maximize well control. Standard safety margins will be allowed to control any pressures that are higher than anticipated. A Gas Emission Response Manual (GSRM) for the permit area is currently under development, which details strategies to be applied in the event of a gas emission. These plans and processes will be introduced in the environmental induction process undertaken by all employees.

3.9 REGIONAL SETTING

3.9.1 CLIMATE

The climate of the area where the project is located is given in Table 3.1.

3.9.2 FLORA AND FAUNA

The country is mostly savanna, although the moist and mountainous east supports tropical evergreen and hardwood forests. Trees include teak and mahogany, knobthorn, msasa, and baobab. Among the numerous flowers and shrubs are hibiscus, spider lily, leonotus, cassia, tree wisteria, and dombeya. The main crops that are grown in the Copper Queen area are cotton and maize. There are around 350 species of mammals,

TABLE 3.1

Copper Queen Climate for 2019

Month	Jan	Feb	Mar	Apr	May	Jun	Jul	Aug	Sep	Oct	Nov	Dec	Year
Average high (°C)	27	26	26	25	23	20	21	23	27	29	28	27	26.2
Average low (°C)	15	15	14	11	8	5	5	6	10	13	15	15	11
Precipitation (mm)	112	130	60	23	8	3	0	0	8	28	99	157	678

including elephants, that can be found in Nembudziya Copper Queen area. There are also many snakes and lizards, more than 500 bird species, and 131 fish species.

3.10 POTENTIAL ENVIRONMENTAL IMPACTS AND MANAGEMENT

It is the philosophy of the interested company to manage environmental risks by removing or mitigating them during the prospecting drilling phase. Table 3.2 describes the potential impacts likely to be associated with the proposed drilling and the unlikely event of methane gas emissions. Each of the effects is discussed in terms of the source, characteristic, potential environmental effect, and management. Additionally, Table 3.2 details "Additional Investigations" that will be carried out as part of the preparation of the EIS to demonstrate whether unavoidable impacts are found to be environmentally acceptable. Based on Table 3.2, the main environmental issues associated with the drilling project are:

i. The effects of drilling cuttings on organisms in the close environment
ii. The effects of drilling noise and operations
iii. The fate (transport and weathering) of gas emissions in the event of an accidental methane gas emission

Although this report focuses on coal bed gas prospecting, it is perhaps important to outline the potential impacts of its extraction. CBM wells are connected by a network of roads, pipelines, and compressor stations. Over time, wells may be spaced more closely in order to extract the remaining methane. Additionally, the produced water may contain undesirable concentrations of dissolved substances. Depending on aquifer connectivity, water withdrawal may depress aquifers over a large area and affect groundwater flows.

The environmental impacts of CBM development are considered by various governing bodies during the permitting process and operation, which provide opportunities for public comment and intervention. Operators are required to obtain building permits for roads, pipelines, and structures; obtain wastewater (produced water) discharge permits; and prepare quarterly reports. As with other natural resource utilization activities, the application and effectiveness of environmental laws, regulation, and enforcement is critical. Violations of applicable laws and regulations are addressed through regulatory bodies and criminal and civil judicial proceedings.

3.11 STAKEHOLDER CONSULTATION PROGRAM

The public engagement through consultation must be done in accordance with EMA requirements. The process ensures that all interested and affected parties (I and APs) are consulted, and their opinions and concerns are documented. Special attention was given to articulating the drilling program regarding the potential positive and negative impacts posed by the exercise.

The process of EIA included the scoping phase and the specialist studies phase. In the scoping phase, the potential impacts were outlined. The specialist studies phase included the detailed analysis of the potential impacts from various perspectives. Both

TABLE 3.2
Potential Impacts, Management and Additional Investigations of CBM Exploration

Source of Risk	Potential Environmental Effects	Mitigating Factors and Management Controls
Physical Disturbances		
Noise/vibration caused by drilling	Disturbance to flora and fauna	Proposed drilling area is not in a known feeding, breeding, or aggregation area nor in a confined area where fauna could be "trapped" when disturbed by noise
		Expected noise levels (McCauley, 2000) are below the guideline threshold, the level of noise that may cause interference to animals of 150 db
		The dominant frequencies of drilling are below the hearing range of animals (100–700 Hz)
		Drilling noise frequencies and intensities (McCauley, 2000) are not in the most sensitive range for animals
		The relatively constant noise source of drilling is less likely to traumatize fauna than erratic sources
		Generally, the drilling instruments will be mounted with noise buffering mechanism
Noise caused by drill rigs, support and supply platform	Disturbance to fauna	Proposed drilling is not in known feeding, breeding or aggregation areas for fauna
		Supply vehicles are well serviced and will not make any noise which can cause interference
Light from drill rigs	Disturbance to animals	Lighting may result in a short-term abundance of some species. The proposed drilling program is of short duration and will not cause irreversible fauna behavior
		Proposed drilling is not in known feeding, breeding or aggregation areas for fauna
Waste Discharges		
Drilling fluids	Disturbance to sensitive environment	Low toxicity drilling fluids are to be used, comprising water and PHG sweeps for the top section holes and a PHPA water-based gel with KCl for the bottom sections
	Adverse effects on water quality	Drilling fluids are reviewed and selected based on technical suitability and by having a minimum overall effect on environment (including ecotoxicity and dosing requirement characteristics)
	Smothering of organisms	Use of drilling chemicals will be minimized as far as is practicable
		High dilution rates – dilutions of >1:100 within 20 m of the discharge are predicted during the drilling phase
		Bunded drill floor catches any chemical spills

(Continued)

TABLE 3.2 (CONTINUED)
Potential Impacts, Management and Additional Investigations of CBM Exploration

Source of Risk	Potential Environmental Effects	Mitigating Factors and Management Controls
Drill cuttings	Disturbance to fauna	Proposed drilling not in known feeding, breeding, or aggregation areas for fauna. The drilling program is of short duration
		Turbid plumes are minimized by cuttings shakers equipment aboard the rig
Drill cuttings	Smothering of organisms	Burrowing invertebrates will tolerate moderate levels of disturbance and are likely to be recolonized in the short to medium term
Laboratory wastes	Localized adverse effects on water quality	Oil-soluble chemicals will be disposed to the oil storage holds
		Other laboratory chemicals will be disposed of in hazardous waste containers for transport for disposal to approved facilities
Cooling water	Localized adverse effects on water quality	High dilution rates would mean that no change in water quality would be detectable outside a localized area
		Cooling water will be discharged at less than 18°C above ambient temperature
Reject water	Localized adverse effects on water quality	Small volumes discharged and high dilution rates would mean that no change in water quality would be detectable outside a localized area
Waste oil and chemicals	Localized adverse effects on water quality	Waste oil and chemicals will be stored in containers and transferred to the authorized disposal sites for disposal
Sewerage, gray waters and putrescible wastes	Localized adverse effects on water quality Nutrient enrichment and bio stimulation of the water Column surrounding the drill rigs	No sensitive resources are known to occur in immediate vicinity of proposed drilling area
		The estimated volume of sewage water produced is 5 L/person/day
		Sewage effluent on the site will be treated in an extended aeration system and comminuted to pass through a screen of less than 25 mm diameter prior to discharge.
		The small volumes of sewage water discharged ensure that only a localized area would be affected by the waste discharge.
		The estimated volume of gray water produced is 10 L/person/day. The small volumes of gray water discharged ensure that only a localized area would be affected by domestic waste discharge
		Domestic wastes discharged after being macerated to a size less than 25 mm

(Continued)

TABLE 3.2 (CONTINUED)

Potential Impacts, Management and Additional Investigations of CBM Exploration

Source of Risk	Potential Environmental Effects	Mitigating Factors and Management Controls
		The gray water is comprised of potable water, soap, and detergents, so none of the identified components of gray water are inherently toxic
Displacement fluid discharge	Leakage of fluids to environment	KCl brine will be used as the displacement fluid during drilling
Solid and Hazardous s Wastes		
General rubbish	Localized adverse effects on environmental quality	All waste material will be kept and disposed of on authorized areas
		Induction of all personnel
		Good housekeeping practices
Hazardous wastes	Localized adverse effects on water quality	No discharge of solid and/or hazardous wastes from the drill rigs
		Wastes stored in appropriate containers
		Hazardous wastes will be labelled and transported to hazardous waste disposal place, in accordance with the Material Safety Data Sheet instructions
		Induction of all personnel includes information on waste management procedures
Atmospheric Emissions		
Power generation	Localized effect on air quality	Fuel usage will be minimized through the sound maintenance of all engines
	Global contribution to greenhouse gases	
Emergency flaring	Localized effect on air quality	Gas will be flared only when hydrocarbons are encountered unexpectedly during drilling
	Global contribution to greenhouse gases	Equipment will be maintained to a high standard to minimize smoke generation
Ozone depleting substances	Localized effect on air quality	Ozone depleting substances will not be used
Physical Presence causing Social Disturbances		
Interference with commercial/ traditional farming	Disruption to farming activities	No commercial farming occurs in the vicinity of the proposed drilling.
		A temporary exclusion zone will apply around the drilling area
Oil, Fuel and Chemical Spill		
Loss of well control and blow-out	Oil spill	Development of comprehensive safety case and validation auditing for the *drill rig* prior to arrival in permit area
		Test the BOP prior to commencement of operations and regularly during operations

(Continued)

TABLE 3.2 (CONTINUED)

Potential Impacts, Management and Additional Investigations of CBM Exploration

Source of Risk	Potential Environmental Effects	Mitigating Factors and Management Controls
		Pressure test casing strings
		Continuously monitor for abnormal pressure parameters during drilling
		Ensure the drill crew is fully trained in emergency well control and gas emission response procedures
Leak from fittings and connections	Oil or chemical spill	Pressure tested equipment
		Pressure low switch on flow lines
Refuelling incident	Localized adverse effects on water quality, toxic effects to biota	Refuelling will be undertaken only during periods of daylight hours
		Transfer hoses will be fitted with "dry break" couplings
		Refuelling operations will be overseen by supervisors
Leaks of hydraulic fluids	Localized adverse effects on environment, toxic effects to biota	Preventative maintenance
		Manned operation (visual detection of release)
Chemical spills	Localized adverse effects on environment, toxic effects to biota	Transfers will be undertaken only during periods of daylight hours
		Transfer operations will be overseen by supervisors
Environmental Impact Assessment at Decommissioning of Prospecting Activities		
Socioeconomic impacts	Loss of employment and loss of income	People will be given recommendations letters showing their experience with drilling work
Impact of road infrastructure	Roads overused and lost shape	Roads improvement exercise will be implemented

phases included public participation processes (PPP). Based on this, the scoping and specialist studies are documented together to keep and ensure the flow of information.

In undertaking the public participation process during the scoping phase and specialist studies, various ways were used to engage stakeholders. The objectives of the PPP in these phases of the EIA process are: to ensure that all potential impacts are identified and documented; to present the findings of the investigations to the interested and affected parties (I and APs); and to provide them with an opportunity to comment on the findings. This is coupled by the need to provide land planners with detailed information on the environmental conservation plans and the legalities associated with them.

3.11.1 METHODOLOGY

Based on the community characteristics and nature of the project, different methodologies were used for stakeholder consultation. Where possible and within the required statutory frameworks, it was also found desirable to structure the process in such a way that it would address the needs and interests of the stakeholders. With regard to the prospecting EIA for the CBM drilling, structured consultation forms were used. The consultant worked very closely with the DA to develop an appropriate program of stakeholder involvement in the Copper Queen area. After meeting to determine the most suitable method of stakeholder engagement, the DA proposed that meetings be convened with the I and APs separately, in order to increase the participation rate and depth of information.

Phase 1: Identification of stakeholders

In the first phase of the engagement process, an initial meeting was held with the key stakeholders (DA, ZRP, Lands office), wherein the milestones for the project were discussed and relevant stakeholders identified. The influence and importance of the support of the ward councillors were discussed as well.

Phase 2: Introducing the stakeholders

The project was introduced to all the interested and affected parties in Copper Queen. The I and APs include the Zimbabwe Republic Police (ZRP), Forestry Commission, Ministry of Transport, President's office, Zimbabwe National Water Authority (ZINWA), and residents, among others. The aim of the project as well as the EIA was explained and major issues of concern were obtained. The importance of community involvement was explained, and verbal recommendations were made by the councillors.

3.11.2 DISSEMINATION OF INFORMATION

The methods explained below were used to make information available for the stakeholder engagement process. The project study report was produced in the first phase of the project and was made available to all the key stakeholders who are I and APs. This is a short report updating the communities of the presence of the company interested in CBM exploration in the area and its intentions. This report was to provide the parties with all the necessary background information regarding the project, such as the current situation regarding mining projects and the need for the new developments. Furthermore, information regarding the EIA and its new regulations was given, which also includes the need and the procedures for a registration for the well-drilling project. Milestones and time frames for the different phases of the project were presented as well.

3.11.3 STAKEHOLDERS

The stakeholder engagement is an integral requirement of the National Environmental Management Act (Act 107 of 1998), the Environmental Conservation Act (Act 73 of 1989), the National Water Act (Act 36 of 1998), and the Forest Act (Chapter 19: 05 –Revised 1996). The process of stakeholder engagement requires the use of various means of information acquisition to comply with legislative requirements and complete the Environmental Impact Assessment the Stakeholder Engagement Process is a very crucial element EIA. The exercise took into consideration communities with the potential to be affected, as well as governmental and non-governmental bodies, which are mandated with the responsibility of ensuring that all possible opinions are taken into consideration in implementing an EIA. The objectives of this process are to identify key issues of concern and possible solutions; access relevant local and traditional knowledge; and ensure that environmental considerations are taken into account in the planning, design, and decision-making of the project.

3.11.4 Approach Used

In undertaking the public participation process, all known and relevant facts pertaining to the proposed project were made available to registered and identified I and APs so that they could participate in a meaningful manner. The approach included:

i. Site visits with technical personnel and I and APs to establish the extent of the study area and the alignment with proposed development
ii. Baseline information survey
iii. An investigation of the site, with particular attention to the type and condition of the veld, potential impact on rare or endangered species, requirements for bush clearing, and potential alternative sites
iv. On-going technical liaising with relevant local officials and the project facilitators regarding the proposed development
v. Preparation of a project study report (PSR) for circulation to I and APs
vi. Ideintification of potential I and APs during discussions with the project facilitators
vii. Provision of written notice to organs of the state (ZINWA, ZESA, AGRITEX, ZRP, Forestry Commission, National Parks and Wildlife, National Museums and Monuments, Local Authorities, President's Office) having jurisdiction over the proposal
viii. Provision of written notice to NGOs and CBOs who might have an interest in the proposal
ix. Individual one-on-one visits, phone calls, and emails to landowners and other I and APs to determine the issues at hand and the perceptions of the public toward the proposal and the current situation
x. Telephone and email conversations with the affected landowners (bid document, covering letter, and project map)
xi. Distribution of draft Environmental Scoping Report to all I and APs for perusal and commenting

3.12 CONCLUSION AND RECOMMENDATIONS

3.12.1 Conclusion

Coal bed gas is a potential source of hazard if it is not removed from underground. In the event of a fire accessing underground deposits, explosions may result, destroying infrastructure and killing people. Such hazards can only be mitigated by tapping the reserves and using them for power generation. CBM is an environmentally friendly resource that can be used for power generation, whereas the escape of unburnt methane into the environment has a high impact on climate change that is over 25 times more potent than carbon dioxide. It is therefore advantageous to burn it into carbon dioxide and reduce its potential negative impact on the environment. Extraction of underground methane gas has very limited environmental impact: the extraction is done over a small area as opposed to coal mining; and all other activities on the land surface are not affected and can even benefit from the power generated. In light of this reasoning, CBM extraction is a good development.

3.12.2 RECOMMENDATIONS

The availability of CBM provides an alternative source of energy and raw material for fine-chemical production, including fertilizers and polymers. For CBM to be extracted, sustainable mining practices need to be considered. The EIA and the EMP then become very critical and must be adhered to in all the CBM exploration phases.

REFERENCES

APASA (2001). Modeling the Expected Concentrations of Total Hydrocarbons Found Within the Woollybutt Plume. *Unpublished Report Prepared by APASA for Eni, Perth, Western Australia*, November 2001.

IUCN (2009). IUCN Red List of Threatened Species. Version 2009.2.

McCauley, R. D., Fewtrell, J., Duncan, A. J., Jenner, C., Jenner, M-N., Penrose, J. D., Prince, R. I. T., Adhitya, A., Murdoch, J. and McCabe, K. (2000). Marine Seismic Surveys – A Study of Environmental Implications. *APPEA Journal* 2000, 692.

Renewable Natural Gas? The Discovery of Active Methane Biogenesis in Coalbeds. (2004). Green Car Congress, 11–16.

Swan, J. M., Neff, J. M. and Young, P. C. (Eds.) (1994). *Environmental Implications of Offshore Oil and Gas Development in Australia, the Findings of an Independent Scientific Review*. Australian Petroleum Exploration Association (APEA), Energy Research and Development Corporation (ERDC), Australia.

Wijffels, S. E., Bray, N., Hautala, S., Meyers, G. and Morawitz, W. M. L. (1996). The WOCE Indonesian Throughflow Repeat Hydrography Sections: 110 and IR6. WOCE, 24.

4 Environmental Impact Assessment for a Solid Waste Disposal Landfill

4.1 INTRODUCTION

The development of sanitary landfills is becoming key in developing countries as a waste management and resource recovery technique. The development of an environmental impact assessment (EIA) report for a new waste disposal facility (landfill) is therefore critical. The proposed landfill development includes but is not limited to the following infrastructure:

i. Fencing of the area
ii. Cell construction
iii. Construction of offices, storage, security, and ablution facilities
iv. Site monitoring system
v. Leachate management
vi. Weigh bridge
vii. Evaporation ponds
viii. Road construction within the fenced area
ix. Incinerator
x. Road construction from residential area to disposal site

In order to achieve authorization for the particular land to become a waste disposal facility, an assessment of the proposed development was carried out.

The Waste and Solid Waste Disposal Regulations, Statutory Instrument No. 6 of 2007 regulates the disposal of effluent and solid waste. It prohibits any person from disposing of waste into a public stream or groundwater without a license. Furthermore, every generator of waste (except households) is now required to come up with a waste management plan by 31 December of each year which deals with quantity and components of the waste, goals for reduction of the quantity and pollutant discharges of the waste, transportation and disposal of the waste, and adoption of environmentally sound management of the wastes. It is an offense for any waste generator to fail to produce a waste management plan.

The proposed landfill is an identified activity in terms of the Environmental Management Act (Waste and Solid Waste Disposal Regulations, Statutory Instrument No. 6 of 2007 deals with regulation of the EIA process and protection of ecosystems). Part 11 of the Act provides that no industrial project shall be implemented unless an EIA has been completed (First Schedule):

"The construction of facilities or infrastructure, including associated structures or infrastructure, for

(l) The construction of a 'BLUE Class' landfill" requires a full EIA, since the proposed development would finally result in a significant change in land use from open field (veld) land to industrial use.

According to EMA regulations these activities may not commence without environmental impact assessment and authorization from EMA and Director-General in respect of which the investigation, assessment, and statement of potential impact of activities must follow the procedure as described in Environmental Management Act (Waste and Solid Waste Disposal Regulations, Statutory Instrument No. 6 of 2007) which deals with regulation of the EIA process and protection of ecosystems.

4.2 TERMS OF REFERENCE

The Terms of Reference (TOR) as approved by the Zimbabwe EMA is described below. The proposed development includes:

 i. Determine the typical composition and character of the general solid waste
 ii. Develop screening criteria for site selection of waste landfill and its associated facilities including composting area, incineration facility, and material recovery area
 iii. Conduct stakeholder consultations including public, civil society, and government on site selection
 iv. Use the screening criteria to develop exclusionary zones for the general waste landfill
 v. Use the screening criteria, maps, interviews, field visits and propose alternatives sites for the general landfill
 vi. Conduct a public and government participation process to reach consensus on one site which will be studied further for the siting of the landfill
 vii. Verify suitability of the screened sites, based on available data, as well as field investigations of geologic, geophysical, hydrogeological, soils, meteorological/atmospheric, and topographic conditions
 viii. Verify that the available sites have at least 20 years of capacity and that the area can access roads and can accommodate projected traffic related to operations
 ix. Produce a general solid waste landfill design acceptable to EMA and other relevant stakeholders
 x. Produce a bill of quantities consistent with the produced general waste landfill design
 xi. Conduct environmental and social impact assessments for the general waste landfill

4.3 SCOPE OF WORK

Following the completion of the specialist investigations, this EIA report was written to present the findings and recommendations according to the Environmental Management Act Chapter 20:27 of 2003. The scope of work included: closure plan of the existing dumpsite, generation of a recycling plan, construction of a new landfill site, and closure plan of the proposed new landfill.

In order to provide scientifically sound information with regard to the potential negative impacts identified, eight specialist studies must be undertaken. An integrated approach was adopted to consider the direct, secondary, and cumulative impacts wherever possible. Each specialist assessed the potential impacts from their specific area of expertise. Subsequently, the findings were integrated and assessed to provide a broad, all-inclusive understanding of the impacts of this project.

4.4 LEGISLATION AND GUIDELINES

The environmental impact assessment was carried out in line with EMA regulations, specifically the Waste and Solid Waste Disposal Regulations, Statutory Instrument No. 6 of 2007.

The aforementioned also specifies and identifies the types of activities which require an EIA process before construction (EMA, Chapter 20:27).

Statutory Instrument 87 of the Environmental Management Act [Chapter 20:27] 2006, Schedule One, stipulates a number of listed activities that could potentially have some detrimental effects on the environment. This information will also be taken into consideration.

In accordance with legislation, the minimum requirements for landfill construction specify that an EIA must be carried out as part of the landfill permitting process.

These activities are subject to Environmental Impact Assessment as specified and regulated by EMA. In addition to the Environmental Management Act (Statutory Instrument 6 of 2007, Statutory Instrument 10 of 2007, and Statutory Instrument 12 of 2007), Environmental Conservation Act, and the EIA regulations, the following laws and regulations have been considered in the assessment process:

4.4.1 CONSTITUTION OF THE REPUBLIC OF ZIMBABWE, ACT NO. 108

Rights of owners and rights of neighbors in respect of environmental quality and health. From this perspective, the mandated municipality is obligated to uphold the public right by ensuring that the environment is kept clean by providing adequate waste management service.

4.4.2 NATIONAL MUSEUMS AND MONUMENTS ACT

Protection and management of heritage resources – The landfill site is located about 2 km from a cemetery; protection from negative impact must be ensured. Furthermore, during construction, if anything of cultural or national value is unearthed, the finding must be reported in time and right action taken by the responsible authority.

4.4.3 NATIONAL HEALTH ACT

During construction, operation, and decommissioning activities, the workers, environment, and public should not be at risk in any way. A healthy environment should be ensured by implementation of appropriate environmental health practices commensurate with the aspect.

4.4.4   FORESTRY ACT CAP 19:05

The conservation of natural vegetation and indigenous species should be upheld throughout the project cycle. Any mutation of the vegetation, probably manifesting through the phenotype should be scrutinized and appropriate action taken to ensure no future complications.

4.4.5   THE CONTROL OF GOODS ACT (CHAPTER 14:05)

Any type of goods which have the potential to cause positive or negative impact should be evaluated and proper recommendations ensured. This prevents pollution by importation activities.

4.4.6   REGIONAL, TOWN AND COUNTRY PLANNING ACT (CHAPTER 29/12)

The town has ensured that all its planning is aligned and there are no future circumstances where non-related infrastructures are found in proximity to each other. The town plans were consulted and appropriate legal procedures followed to ensure that the structure would be on a legally provided land and in compliance with the Act.

4.4.7   ENVIRONMENTAL MANAGEMENT ACT [CHAPTER 20:27] 2006

All specific actions and compliance are to be upheld to ensure public and environmental protection. Specific statutory instruments are to be observed throughout the project cycle.

4.4.8   ENVIRONMENTAL MANAGEMENT ACT (ENVIRONMENTAL IMPACT ASSESSMENT AND ECOSYSTEMS PROTECTION)

The appropriate procedures were taken from project feasibility to construction in line with EMA requirements. The standards will be upheld throughout the project cycle.

4.4.9   NATIONAL BIOTECHNOLOGY AUTHORITY ACT (ACT 3, 2006)

The Act should be upheld by the appropriate authority. Periodic tests should be conducted throughout the project cycle to avoid waste management related harm.

4.4.10   PARKS AND WILDLIFE ACT (CH 20:14)

The area in the vicinity of the site has no park. If any such development were to commence in the interim, inter-relationship should be scrutinized to ensure no negative impacts on the park area.

4.4.11   PROTECTION OF WILD LIFE (INDEMNITY) ACT (CH 20:15)

The area is blessed with a wide diversity of animals. Baboons have invaded the area and are often found feeding at the dumpsite. This leads not only to disease contamination of livestock, who feed in the area and could potentially infect humans, but also to overpopulation of the baboon. Although the baboons in this respect have become a menace, they deserve protection as do domestic animals. Fencing and humane population control are options for reducing this problem.

4.4.12   TOURISM ACT (CH 14:20)

Tourist attraction centers should not be affected by the development. It is important to ensure that no negative impact is exerted on potential tourist attraction centers.

4.4.13   WATER ACT (CH 20:24)

Water is a scarce resource. Wherever a project is pursued, water should be used sustainably and protected at all costs. During operation, appropriate measures must be taken to prevent surface and groundwater pollution. After the EIA process is completed in terms of Section 99, the Director General will issue a certificate in terms of Section 100. The permit application (EIA report) will be submitted to EMA. After completing the EIA and authorization by Director-General in terms of Section (1) of EMA (20:27), the site will be registered under the Environmental Management Act.

4.5   DEFINITION OF KEY WORDS IN LANDFILL MANAGEMENT

4.5.1   SOLID WASTE

Municipal solid waste, commonly known as trash or garbage, refuse, or rubbish and consisting of everyday items that are discarded by the public.

4.5.2   SOLID WASTE DISPOSAL

The final placement of refuse that is not salvaged or recycled.

4.5.3   SOLID WASTE MANAGEMENT

Systematic control of generation, collection, storage, transport, source separation, processing, treatment, recovery, and disposal of solid waste.

4.5.4   DUMPSITE

A place approved by the municipality for the discarding or dumping of waste.

4.5.5   LANDFILL

A disposal site, also called a sanitary landfill, is where solid waste such as paper, glass, and metal, is buried between layers of dirt and other materials in such a way as to reduce contamination of the surrounding land. Modern landfills are often lined with layers of absorbent material and sheets of plastic to keep pollutants from leaking into the soil and water.

4.6   PROJECT ACTIVITIES

The project activities included development of an EIA report on the prospective site of a new solid waste disposal facility, analysis of the current dumpsite to determine its closure, and creation of recycling, monitoring, and closure plans for the new solid waste disposal facility.

The proposed development includes:

i. Closure plan for the existing dumpsite
ii. Fencing of the area
iii. Disposal cell preparation plan
iv. Offices, storage, security, and ablution facilities siting
v. Site monitoring system
vi. Weigh bridge
vii. Leachate management system
viii. Storm drainage facility
ix. Evaporation pond
x. Road construction within the fenced area
xi. Development of a recycling plant
xii. Operation and closure plan of the proposed landfill

In order to achieve authorization for the particular land to become a solid waste disposal facility, a scoping study was conducted to identify the potential impact of the planned development. The exercise was done in accordance with the Statutory Instrument No. 6 of 2007 of the Environmental Management Act (Chapter 20:27). The proposed landfill is an identified activity in terms of the EMA Act (Waste and Solid Waste Disposal Regulations, Statutory Instrument No. 6 of 2007) and deals with regulation of the EIA process and protection of ecosystems. Part 11 of the Act provides that no industrial project shall be implemented without an EIA having been completed (First Schedule). These regulations identify various activities which may have a substantial detrimental effect on the environment. In addition, the regulations list procedures for assessing potential associated environmental impacts.

4.7   NEED AND DESIRABILITY OF THE PROPOSED ACTIVITY

In Zimbabwe, it is a progressive requirement for each municipality to have an Integrated Solid Waste Management Plan (IWMP). The necessity of the disposal site is in line with Waste and Solid Waste Disposal Regulations, Statutory Instrument No. 6 of 2007, which regulate the disposal of effluent and solid waste. It prohibits any person from disposing waste into a public stream or groundwater without a license. Furthermore, every generator of waste (except households) is now required to come up with a waste management plan by 31 December every year which deals with quantity of waste, components of the waste, goals for reduction of the quantity and pollutant discharges of the waste, transportation and disposal of the waste, and adoption of environmentally sound management of the wastes. It is an offense for any waste generator to fail to produce a waste management plan. In line with this requirement, all town councils are planning to develop their IWMP to ensure quality operation of the proposed landfill. The decision to construct a new waste disposal facility was influenced by several factors, such as dwindling capacity of current disposal facilities and sanitary concerns.

The size of the old dumpsite is about 10 hectares, it is not fenced, and a dirt road bisects the site. There is one access way but not a gateway. There is no security guard or signage at the site. Although there is normally one staff from the municipality present, the public often dumps at random. Daily covering of the waste is not practiced and different types of wastes are disposed of in one place. The size of the new facility will be 5 hectares.

Because the town is on the border and attracts tourism, accurate population projections are not possible, so disposal capacity will be based on waste volume statistical trends rather than the household number and population. The town has a permanent population of around 33 000 people and a transit population of 1 million due to tourism and people moving through the Zambian and Botswana border. Various factors have a bearing on both the character and growth of the town population.

4.8 HUMAN POPULATION DYNAMICS

The town will have a great impact on the amount of solid waste generated as well as the size of the landfill to be constructed.

4.9 THE SOCIOECONOMIC BASE

The town has a dualistic economic structure whereby the formal sector enjoys modern patterns of investments, while the informal sector is more indigenous and exhibits greater ties with rural areas. In the formal sector, tourism is the main economic activity, provides most employment opportunities, and generates considerable revenue for the town. Apart from those directly employed in the tourism sector, hundreds of people in the town supplement their incomes by selling locally made handcrafts and artifacts to tourists. The socioeconomic base of the town comprises the following major infrastructural facilities: transport routes, health facilities, industrial establishments, and tourist and accommodation facilities.

4.10 TRANSPORT INFRASTRUCTURE

The provision of adequate transport infrastructure is an important prerequisite without which it would be very difficult to move freight, passengers, and even solid waste from sources of generation to disposal sites. For instance, in informal settlements which may not have any roads, it is practically impossible to provide formal refuse collection services using conventional methods. There are two regional roads namely the Bulawayo and Kazungula highways that connect the town to other urban centers in Zimbabwe and neighboring countries, respectively. Apart from these major roads, the town has numerous link roads. While the road network is fairly developed, a major cause for concern is that few roads are all-weather and the rest are either gravel or earth roads. As of December 2002, the municipal dump was connected to the rest of the town by a gravel road, which gets slippery during the rainy season, thereby slowing the speed of refuse trucks. Apart from the road network described above, the town is linked to Bulawayo (the second largest city in Zimbabwe) and Livingstone (a resort town in Zambia) by a railway line.

4.11   HOUSING FACILITIES

There are three basic housing units providing shelter to residents of the town. Data provided by the Department of Housing show that as of December 2002 there were 8 819 housing units in the town. About 60% of these housing units are classified as informal, while 30% are classified as formal. The remainder are considered semi-formal.

Formal housing units are those permanent structures, built of brick with asbestos or tiled roofs. Most of them have basic internal water and sanitary systems. These housing structures constitute the bulk of housing units in the formal residential areas and flats houses located mostly in the town center. Semi-formal housing units are simple shelters erected by private companies as a stop-gap measure to accommodate their employees considered essential to their business but for whom no permanent houses are available. Such housing structures are found in Gardini, Zimsun, Savanna, and Zimcast residential areas. As they have, over a period of time, attained a degree of quasi-permanence, the municipality provides basic services, such as waste collection. The third and largest category of housing units comprises informal structures erected mostly using cheap and readily available materials such as plastic, scrap metal, and even cardboard boxes some of which are recovered from waste dumps scattered in the town. These structures are built without any plan and have no legal status.

4.12   HEALTH FACILITIES

There are three major health facilities offering a fairly wide range of medical services. These include: (i) The General Hospital that is run by the Ministry of Health and Child Welfare (MoHCW), (ii) a local municipality clinic and (iii) 8 privately owned surgeries and pharmacies. These health facilities are struggling to meet the health needs of the growing population, given that they were established to cater to 8 000 residents. The medical waste generated at these facilities is incinerated.

4.13   ECONOMIC OPPORTUNITIES

In 1995 employment data showed that 51.5% of the population were engaged in formal employment and 13.5% were involved in informal activity, while 35% were classified as unemployed. Although up-to-date employment records are difficult to obtain, the 1992 national population census showed that 23.98% of the economically active were classified as unemployed.

4.14 WASTE MANAGEMENT SYSTEM

Currently the town has a waste management system supported by a few vehicles. One 7-ton municipal garbage truck comes into the dumpsite every day. On average, each frequents the dumpsite 3–4 times a day. Several individual trucks also come in with a variety of general waste. There are about 22 general hands, 2 supervisors, and 2 professionals who patronize the dumpsite regularly and a couple of other day trippers who come in on an irregular basis. The patrons search through the waste for valuables, making them vulnerable to contagious diseases. The waste is collected daily in the industrial areas, twice a week in the low-density, and once a week in the high-density suburbs. Approximately, 28 tons of general solid waste described in Table 4.1 is collected per day.

4.15 WASTE SEGREGATION

The public is not allowed to enter the disposal site due to the risk of disease and danger from operating machinery and is supposed to be fenced to deter people and animals from entering the premises. This also ensures that people and animals do not salvage the valuables and edibles from the disposals site.

It is well understood that not all waste is waste, as some can be reused or recycled according to Table 4.1 which indicates the types of waste found at the dumpsite. The principle of waste separation at sources is widely practiced in Europe, America, and other developed countries to ensure that valuables and edible waste are separated or salvaged before being mixed with the useless waste. This approach reduces the risk of valuable waste stream contamination and increases the salvaging capacity by the people.

In practice, segregation is the sorting of waste to make sure the right waste goes to the right place, that is, what is to be recycled, what goes to the landfill, and what needs special handling, like anti-freeze. Waste segregation requires the cooperation of the citizens and businesses for it to work. Often three color-coded bags or bins are put on a collection site. Each one is for a specific type of waste, such as organic waste, non-recyclables, and recyclables.

TABLE 4.1
The Municipal Waste Composition (Masocha, 2004)

Waste Category	Composition (%)
Paper and cardboard	34
Organic wastes	26
Plastics	15
Glass and ceramics	6
Metals	5
Clothes and textiles	4
Leather and rubber	2
Miscellaneous/inert materials	8

Waste segregation is the best solution to garbage woes and requires land space for composting of the biodegradable waste.

The 3-R policy – Reduce, Re-use, and Recycle leads to segregation of degradable and non-degradable waste. However, in the town at the moment, solid wastes are being disposed of either by the formal route i.e. to the designated dumpsite and informally on illegal dumpsites.

4.16   ENVIRONMENTS THAT MAY BE AFFECTED (BASE LINE STUDY)

4.16.1   TOPOGRAPHY

The town and the surrounding rainforest are preserved as a 23.4-kilometer national park and form one of Zimbabwe's four world heritage sites. The Zimbabwe Parks and National Wildlife Management Authority has maintained the town and the surrounding rainforest virtually as they were when Livingstone first saw them almost 140 years ago.

4.16.2   CLIMATE

The weather is mainly divided into summer, rainy, and winter seasons. Winter runs from May to mid-August and is mild and dry. During this time the temperatures range from highs of 25–27°C to lows of 7–10°C. The summer is from mid-August to the end of April and is generally hot to very hot and wet during the rainy season. The rainy season runs from mid-November to April. Average temperature ranges from highs of 32–34°C to lows of 15–19°C. From the end of August, the temperatures start rising, and from September through to November, it is very hot and dry. At this time, the typical African landscape turns very brown and can feel a bit harsh.

The bird life is superb all year round, but in September it is exceptional with the arrival of numerous migrants. Among these are the spectacular carmine bee-eaters arriving en masse and nesting in the sand cliffs along the Zambezi River. November is when you can expect the rains to start. It becomes hot, wet, and quite muggy. The area does not experience rains day after day, but rather thunderstorms that build up in the late afternoon, followed by a torrential downpour which is over in about an hour. Although some very exciting electrical storms can be experienced, the rains make some of the roads in the national parks inaccessible, resulting in park closures. The rainy season normally lasts until about March/April. The climatic data for the town is described in Table 4.2.

4.16.3   GEOLOGY

The recent geological history of the town can be seen in the form of the gorges below the falls. The basalt plateau over which the Upper Zambezi flows has many large cracks filled with weaker sandstone. In the area of the current falls the largest cracks run roughly east to west (some run nearly northeast to southwest), with smaller north-south cracks connecting them.

TABLE 4.2
Climatic Data for The Town (Weather Source, 2011)

Month	Daily Average Maximum (°C)	Daily Average Minimum (°C)	Mean Total Rainfall (mm)	Mean Number of Rain days
Jan	30	18	168	14
Feb	29	18	126	10
March	30	17	70	7
April	29	14	24	2
May	27	10	3	1
June	25	6	1	0
Jul	25	6	0	0
Aug	28	8	0	0
Sept	32	13	2	1
Oct	33	17	27	4
Nov	32	18	64	8
Dec	30	18	174	13

Over at least 100 000 years, the falls have been receding upstream through the Batoka Gorges, eroding the sandstone-filled cracks to form the gorges. The river's course in the current vicinity of the falls is north to south, so it opens up the large east-west cracks across its full width, it then cuts back through a short north-south crack to the next east-west one. The river has fallen in different areas into different chasms which now form a series of sharply zigzagging gorges downstream from the falls.

Apart from some dry sections, the Second to Fifth and the Songwe Gorges each represents a past site of the falls at a time when they fell into one long straight chasm as they do now. Their sizes indicate that we are not living in the age of the widest-ever falls. The falls have already started cutting back the next major gorge, at the dip in one side of the "Devil's Cataract" (also known as "Leaping Waters") section of the falls. This is not actually a north-south crack, but a large east-northeast line of weakness across the river, where the next full-width falls will eventually form.

4.16.4 Soils

The area is underlain by basalt which is exposed in the area adjacent to the Zambezi River. Most of the land is overlain by Kalahari sands which are gently undulating. The most spectacular feature of the Zambezi River is a 300 m deep gorge cut into the basalt. The site for the future disposal site was analyzed by soil specialist. The objective was to ascertain the compatibility of the soil to the structure to be built. Basically, the clay content is the aspect which is under consideration. Visual and texture analysis reflected that the soil is capable to arrest leachate pollutants into the immediate and adjacent ground.

4.16.5   HYDROLOGY

The drainage of the town is dominated by the Zambezi River with its numerous tributaries, braided valleys, islands, swamps, and dramatic ziz zag gorges, which constitute principal landscape features. The main tributary river, which drains into the Zambezi from Victoria Falls Town is the Masuie river. The Masuie is about 200 m to the south of the current municipal waste dump. Also close to the municipal dump is a stream, which flows into the Masuie. Although the stream is dry for the greater part of the year, there is a danger that in summer it transports contaminants from the dump to the Masuie, which is a cause for environmental concern. The municipality collects samples from the river for quality analysis. The latest analyses are reflected in Table 4.3.

4.16.6   FLORA AND FAUNA

Mopane woodland savannah predominates in the area, with smaller areas of Miombo and Rhodesian teak woodland and scrubland savannah. Riverine forest with palm trees lines the banks and islands above the falls. The most notable aspect of the area's vegetation, though, is the rainforest nurtured by the spray from the falls, containing plants rare for the area such as pod mahogany, ebony, ivory palm, wild date palm, and a number of creepers and lianas. Vegetation has suffered in recent droughts, and so have the animals that depend on it – particularly antelope.

The national parks have abundant wildlife including sizable populations of elephant, buffalo, giraffe, zebra, and a variety of antelope. Lion and leopard are only occasionally seen. Vervet monkeys and baboons are common. The river above the falls contains large populations of hippopotamus and crocodile. Elephants cross the river in the dry season at particular crossing points.

Although the gorges are mainly known for 35 species of raptors, klipspringers and clawless otters can be glimpsed there as well. The taita falcon, black eagle, peregrine falcon, and augur buzzard breed there. Above the falls, herons, fish eagles, and numerous kinds of waterfowl are common. The river is home to 39 species of fish below the falls and 89 species above it. This illustrates the effectiveness of the falls as a dividing barrier between the upper and lower Zambezi. The old dumpsite is regularly visited by vultures and stock birds. There is a challenge to keep these birds away but the new disposal site will not present access to these birds.

TABLE 4. 3
Masuie River Water Sample Analysis

Aspect	Quantity	Comment
Temperature	31°C	Entropy too high
Turbidity	12.19 NTU	It's above acceptable at around 5 NTU
Total dissolved solids	272 ppm	Higher than expected at around 90 ppm
Conductivity	96.7 mV	Within acceptable limits
Ph	8.55	Not neutral as expected

4.17   SITE SELECTION

4.17.1   INTRODUCTION

For a landfill to be built, the proponent has to make sure that certain steps are followed. In most parts of the world, there are regulations that govern where a landfill can be placed and how it can operate. The whole process begins with someone proposing the landfill.

In Zimbabwe, taking care of trash and building landfills are local government responsibilities. Before a city or other authority can build a landfill, an EIA must be completed on the proposed site to determine:

- The area of land necessary for the landfill
- The composition of the underlying soil and bedrock
- The flow of surface water over the site
- The impact of the proposed landfill on the local environment and wildlife
- The historical or archaeological value of the proposed site

It is required that feasible, reasonable alternatives to the project site are identified, investigated, considered, and evaluated in terms of various factors such as social, economic, and biophysical. Based on the exclusionary approach, the areas in which the potential sites can be located were determined. The town could only avail of an area within the vicinity of the current dumpsite. Besides that, the area identified is in the windward direction, it is the area surveyed for future town expansion. Within the area, three sites were identified. These sites were investigated thoroughly, and the following sections provide brief descriptions of each and their advantages and disadvantages. During the EIA, three sites were identified as candidate sites and were subjected to survey. In determining the most appropriate site a number of criteria were considered – such as economic, environmental, as well as public influenced criteria – and analyzed. Based on these analyses, the most suitable site for the future development was identified.

4.17.2   SITE SELECTION AND ALTERNATIVES RANKING CRITERIA

A ranking approach was used to select the most appropriate disposal site based on the exclusionary approach. Within the area, three different locations were identified for a future development of a waste disposal facility from a theoretical perspective. All the other potential sites were dropped since they were located in the windward direction.

The three potential locations were studied with regard to various aspects which impact on economic, environmental, and pubic criteria, which include:

- Distance – the further from the area of generation the more expensive are the collection/transportation costs
- Size – must be large enough to last until the middle term life span in order to justify the capital expenditure

- Access – new roads would have a high cost and environmental implications
- Availability of soil on site to reduce the cost for the possible provenance of cover material
- Visibility of the site should be as low as possible in order to avoid unnecessary cost implications
- Distance and availability of surface and groundwater need to be as far as possible in order to avoid environmental pollutions e.g. leachate impact
- Distance of the site to settlements needs to be known. A displacement of inhabitants should not be seen as a solution to the problem
- Are there any wind directions disturbing inhabitants in nearby settlements

As Table 4.4 shows, a ranking was done, whereby marks were allocated based on the required criteria. A mark of 5 was the highest achievable mark for a criterion and a site, whilst a mark of 1 indicated the minimum.

As Table 4.4 represents, Site 1 achieved the highest score with regard to the site suitability or good condition. It was therefore recommended to develop Site 1 as the future waste disposal facility.

4.18   WASTE CLASSES

Based on the "Minimum Requirements for Waste Disposal by Disposal Site" (South African Standards, 1998), waste is classified into two types, namely General (G) or

TABLE 4.4
Ranking of the Three Proposed Sites for the Landfill

Candidate Site		Site 1	Site 2	Site 3
Economic Criteria	Distance	2	5	2
	Size	3	5	3
	Access	3	5	3
Environmental Criteria	Groundwater	2	2	2
	Surface Water	5	5	5
	Soil depth	4	4	4
	Setting	5	2	2
Public Acceptance	Distance	5	1	2
Criteria	Visibility	5	1	3
	Wind	5	4	3
	Total Score	49	34	32

Mark	Ranking
1	Unacceptable
2	Bad
3	Acceptable
4	Good
5	Very good

Hazardous (H) waste. General solid waste is a generic term for waste such as domestic, commercial, certain industrial wastes, and builders' rubble which, because of its composition and characteristics, does not pose a significant threat to public health or the environment.

Hazardous waste is waste which, because of its inherent characteristics (e.g. toxicity, corrosivity), can, even in low concentrations, have a significant negative effect on public health and/or the environment. Based on the fact that the proposed waste disposal facility should service all households, whereby household service is the main focus, only general waste should be disposed of at the facility. Based on historical data from the former site, additional investigations, site inspections, and meetings held by municipalities and involved parties, the site should only receive general waste. Therefore, the waste disposal site is classified as a "G" facility.

4.18.1 SIZE OF WASTE STREAM

Apart from determining the disposal site airspace volume required for disposal, i.e. the disposal need, the size of the waste stream determines the size of operation and hence the class of disposal site. The size of a disposal facility is usually influenced by the served population. Along these lines, the rate of waste disposal is calculated in tons per day or by the amount of waste that accumulates over a given period of time, measured in cubic meters of airspace occupied or hectares of land covered. The Legal Minimum Requirements require that the size of the disposal site to be determined by the expected amount of waste being deposited at the end of operation, whereby the average daily rate of deposition is based on a 365-day year. The amount of waste that is currently being generated by the town is around 840 tons/month, or 28 tons/day.

4.18.2 LEACHATE

The potential for a waste disposal site to generate significant leachate must be determined in order to (i) identify the need for leachate management and (ii) classify the site in terms of the Minimum Requirements. The potential for significant leachate generation depends on the site water balance. This is dictated primarily by ambient climatic conditions but also by site-specific factors, such as the moisture content of the incoming waste and/or the ingress of either ground or surface water into the waste body.

As the moisture content of the general waste generated in the study area is low, and since the new disposal site will be properly engineered and operated, significant leachate should not be generated. The disposal site would therefore be classified as a B– site, i.e. no significant leachate generation is anticipated and consequently a leachate management system would not be mandatory. Based on the above information and the Minimum Requirements, the waste disposal site is classified as GMB–, indicating a general solid waste disposal facility of medium size and low leachate generation.

4.19   PROJECTED DISPOSAL SITE AIRSPACE AVAILABILITY AND SITE LIFE

The capacity or expected life, of a waste disposal site can depend on various criteria. Often the airspace requirement is calculated up front, and a suitably sized site is identified. The airspace calculation for the waste disposal site is based on the population as well as the population growth rate over the estimated life span of the future disposal site. The disposal site has to have a capacity to service the entire population of the municipality. The total airspace within a waste disposal facility will vary depending on the depth of excavation, compaction methods, efficiency of operations, and the height above ground that is used for disposal.

Taking into account the allowances for compaction of the waste to be 840 tons/ month and a provision of cover material at a ratio of 1 to 4 (soil to waste) the required airspace over a period of 25 years is 850 000 m³. This estimate takes into account the increase in population growth as well as the amount of waste generated.

The above represents total airspace including the required cover material. This means waste going into the waste disposal site would be 300 000 m³. The calculation above does not include any waste minimization methods or recycling activities. Considering minimum requirements and the need of implementing waste reduction and minimization activities, the life span of the waste disposal site should be much longer. Assuming a 75% waste reduction rate, the above-mentioned airspace of 850 000 m³ should last for 32 years instead of 25 years.

4.20   LANDFILL COVER MATERIAL AND AVAILABILITY

The availability of cover material close to the site is critical. It is also one of the issues why the scoping report covers the determination of the soil condition and soil types at the proposed area. The soil excavated on the site will be used as cover. For the operation of the future waste disposal site a cover-to-waste ratio of 1:4 is recommended within the Minimum Requirements. Thus, the operation of the facility requires that 4m³ of waste is covered with 1 m³ of soil.

With regard to the envisaged volume of generated waste over a period of 25 years (300 000 m³), (including bulking) one hundred and ten cubic meters of soil would be necessary to ensure adequate operation of the waste disposal site. The geo-hydrological survey has established that soil can be excavated down to a depth of, at most, 1.5m.

It is calculated that about 850 000 m³ total airspace is required on an area of about 6 ha. The construction at about ground level can only take place in a 30% angle. At the same time, it is assumed that the excavation can take place up to 1–2 m below ground level. The footprint for the future waste disposal site is 200 m by 160 m. Therefore, if the perimeter is calculated to be 720 m it should be possible to build up the disposal facility to 4.5 m above ground level, assuming it is possible to excavate 3 m below ground level.

4.21   LANDFILL WASTE RECYCLING TECHNIQUES

By introducing separation at source techniques which lead to waste minimization and recycling, the life span of the waste disposal site can be extended. Furthermore, all organic waste can be converted into bio fertilizers through vermicomposting thereby further increasing the lifespan of the landfill.

4.22   PUBLIC PARTICIPATION PROCESS

The process of the EIA included the scoping phase and the specialist studies. Within them, the public participation process (PPP) is done. Based on this, the scoping and specialist studies participation process are documented together to keep and ensure the flow of information. In undertaking the public participation process, during the scoping phase and specialist studies, various ways were used to engage stakeholders. The objective of the PPP in these phases of the EIA process is to present the findings of the investigations to the interested and affected parties (I and APs) and provide them with an opportunity to comment on these. This is coupled with the need to provide land planners with detailed information on environmental conservation and the legalities associated with it.

4.22.1   METHODOLOGY

Based on the community characteristics and nature of the project, different methodologies can be used for stakeholder consultation. Where possible and within the required statutory frameworks, it is also desirable to structure the process in such a way that it would address the needs and interests of the stakeholders. With regard to the EIA for the new waste disposal facility the following public consultation techniques were used:

   i. Consultation forms
   ii. Media

The consultant worked very closely with the municipality to develop an appropriate program of stakeholder involvement. Consultative meetings were held with the municipality to determine the most suitable method of stakeholder engagement. The municipality suggested having two community meetings, each at separate places, in order to increase the participation rate.

Phase 1: Identification of Stakeholders

In the first phase of the engagement process an initial meeting was held with the key stakeholders and potential partners. At this meeting the different milestones for the project were discussed and relevant stakeholders identified.

The impact and importance of the support of the ward councillors were discussed as well.

Phase 2: Introducing the stakeholders

The project was introduced to all the interested and affected parties in including the Zimbabwe Republic Police, forestry commission, airport staff, Zimbabwe Tourism Authority, Ministry of Transport, President's Office, ZIMRA, Immigration, ZINWA, Residents, National Parks, Green Fund, Go Green, Environment Africa, and African Recycling.

The aims of the project and the EIA were explained and major issues of concern were obtained. The importance of community involvement was explained and recommendations were made by the councillors.

Phase 3: Stakeholder Meetings and Public Meetings

The third level of stakeholder engagement was through stakeholder and public meetings. During the meeting:

i. The project phase was introduced
ii. Summary and findings so far were presented
iii. A discussion was held where all stakeholders were asked to raise their issues, concerns, and questions
iv. Questions raised were answered by the client and the consultant.
v. All issues were captured during the meeting

A representative from the municipality was available at the meeting. This ensured a direct communication between the municipality and residents. This was a good sign of transparency and support.

4.22.2 DISSEMINATION OF INFORMATION

The methods listed below were used to make information available for the stakeholder engagement process.

4.22.3 PROJECT STUDY REPORT

The project study report was produced in the first phase of the project and was made available to all the key stakeholders who are interested and affected parties. The purpose of this report was to provide the parties with all the necessary background information regarding the project as e.g. the current situation regarding waste management, the need for the new development, and, therefore, the project. Furthermore, information regarding the EIA and its new regulation was given, which also includes the need and the procedure of a registration for the new waste disposal facility. Different milestones and timeframes for the different phases of the project were given.

4.22.4 NEWSPAPER ADVERT

To access as many people as possible, one newspaper was selected where the advertisement would be published.

4.22.5 NOTICE TO STAKEHOLDERS AND AFFECTED PARTIES

In addition to the above-mentioned methods, interested and affected parties were informed via letter and e-mail with support of the municipality. This was done to ensure everyone was involved in the consultation process.

4.22.6 STAKEHOLDERS

The stakeholder engagement is an integral requirement of the National Environmental Management Act (Act 107 of 1998), the Environmental Conservation Act (Act 73 of 1989), the National Water Act (Act 36 of 1998), and Forest Act (Chapter 19: 05 –Revised 1996). This process requires the uses of various techniques to acquire adequate information which take legislation into account.

For the accomplishment of an EIA, the stakeholder engagement process becomes a very important element. It included affected communities, governmental, and non-governmental bodies, which are required to make sure that all possible views are taken into consideration by implementing an EIA. The objectives of this process are to identify key issues of concern, offer possible solutions, and access relevant local and traditional knowledge. This process ensures that environmental considerations are taken into account in the planning, design, and decision-making of the project.

4.22.7 APPROACH USED

In undertaking the public participation process, all known, relevant facts pertaining to the proposed project were made available to registered and identified interested and affected parties (I and APs) so that they could participate in a meaningful manner.

The approach included:

i. Site visits with technical personnel and I and APs, to establish the extent of the study area and alignment with proposed development
ii. Baseline information survey
iii. An investigation of the site, with particular attention to the type and condition of the veld, potential impact on rare or endangered species, requirements for bush clearing, and potential alternative site.
iv. On-going technical liaising with relevant local municipal officials and the project facilitators regarding the proposed development
v. Preparing a project study report (PSR) for circulation to I and APs
vi. Identifying potential I and APs during discussions with the project facilitators
vii. Giving written notice to organs of the state (ZINWA, ZESA, ZIMRA, Immigration, ZRP, Forestry Commission, local authorities, President's Office) having jurisdiction over the proposal
viii. Giving written notice to non-governmental organizations (NGOs) and community-based organizations (CBOs), etc., who might have an interest in the proposal
ix. Conducting one-on-one meetings/visits, phone calls, and emails with landowners and other I and APs, to determine the issues at hand and the perceptions of the public toward the proposal and the current situation
x. Telephonic conversations with the affected landowners, as well as email correspondence (bid document, covering letter and project map)

    xi. Several individual meetings with any concerned landowners and other I and APs

    xii. Advertisements in the local press informing the public of the proposed disposal facility site

    xiii. Preparation of draft environmental scoping report as well as submission of copies were made available to all I and APs for perusal and commenting

4.22.8   Discussion

The purposes of the stakeholder engagement process were: to inform and introduce the project to the community in which it is going to take place; to meet and register those who would like to be regarded as interested and affected parties; to obtain the views, issues, and concerns of the I and APs about the project. Besides the concerns raised by the community, it was also important to receive feedback from interested and affected parties such as different governmental bodies and NGOs.

4.22.9   Summary of Comments by the Environmental Agent Practitioner

    i. A closure plan for the current dumpsite must be implemented

    ii. All material which is recyclable must be recycled

    iii. There should be lining of the cells by impermeable media or clay soil

    iv. Ensure covering of waste at all times

    v. There must be security personnel

    vi. Recovery of methane gas

    vii. Leachate collection and leachate dams must be in place

    viii. Separation at source techniques must be implemented

    ix. Eco-composting must be implemented from household level

    x. An operating and closure plan for the proposed landfill must be employed

4.23   SPECIALIST STUDIES

This chapter focuses on the three specialist studies which were done and the method which was used by the specialist in undertaking the EIA process. The details of the methods are not part of this EIA report. This approach was taken to avoid large volumes of paper and presentation of too much primary information. An interactive approach among specialists was taken, eliminating the excess information which has no bearing on the report content and project objective. The concerted specialist findings were incorporated in this report.

4.23.1   Steps Undertaken in Accordance with the Plan of Study

Each specialist study was approached in a distinct and appropriate way to determine and access relevant information. Below are the approaches which were used by the specialists:

Avifauna, Fauna, and Vegetation

To carry out the avifauna assessment the following steps were taken: review of literature, data collection, and information analysis. The compiled report included review of results, impact analysis, and mitigation recommendations.

Heritage Assessment

To carry out the heritage assessment, the following steps were taken:

- Review of literature
- Site visits

The compiled report included review of results, impact analysis, and mitigation recommendations.

Geohydrology and Geotechnical Assessment

To carry out the geo-hydrological and geotechnical assessment the following steps were taken: review of literature, site visits, geological survey, hydrogeological survey, geotechnical survey, laboratory testing (water and soil sample tests), map drawing, and data collection and analysis. The compiled report included review of results, impact analysis, and mitigation recommendations.

4.23.2   METHOD USED TO IDENTIFY SIGNIFICANCE BY SPECIALISTS

This section explains how the specialists examined the significance of the potential impacts on the environment; adjudicated the proposed development; and examined the key issues/impacts which may be predicted to occur as a result of the proposed development. The impacts discussed look into all project stages, from site operation phase through to decommissioning and closure of the site because this is an identified site with known use after closure.

Mitigation measures are also discussed. In this context mitigation means to "allay, moderate, palliate, temper, or intensify". Mitigation seeks to find better ways of undertaking activities, to minimize or eliminate negative impacts, and to enhance and maximize positive impacts on people and the environment.

Impacts are described in terms of the impacts' nature, duration, significance, and extent. A positive impact is one that enhances the existing environment, whereas a negative impact is one that degrades the environment. Duration is referred to as short, medium, and long term. An impact of low significance will have only a limited effect on the environment, whereas an impact of high significance will have a major impact on the environment. The definitions outlined below draw from those described in the EIA regulations guideline document (Department of Environmental Affairs and Tourism, 1998 South Africa).

Nature of Impact

Impacts are firstly classified as positive (beneficial) or negative (destructive), due to the nature of the impact.

A positive impact is one which enhances the existing environment

A negative impact is one which degrades the environment.

Extent or Spatial Scale

Site: impact affects the whole or measurable part of the activity area

Localized: site and immediate surrounds, adjacent households/village

Sub-regional: geographic area or municipal scale
Regional: provincial scale or impacts across provincial borders
National: Zimbabwe

International: neighboring countries with respect to shared borders or resources e.g. Zambia, Botswana, and Angola.

Duration

Short term: impact will disappear with mitigation or will be mitigated through natural processes in less than 5 years
Medium term: impact will last 5–10 years, whereafter it will be entirely negated
Long term: impact will last for the entire operational life-of-development but will be mitigated by human intervention or natural processes thereafter
Permanent: non-transitory impacts that cannot be mitigated by man or natural processes

Intensity/Severity

This is a relative evaluation of all activities that describe the degree of destructiveness of an impact, whether it destroys the impacted environment or alters its functioning.

No effect: neither systems nor parties not affected may not be possible to determine
Low: impact alters the environment in such a way that the natural processes or functions are not affected
Medium: affected environment is altered, but function and process continue although in a modified way
High: function or process of the affected environment is disturbed to the extent that it temporarily or permanently ceases or constitutes a safety hazard

The assessment of the effects of an impact hereunder assumes that mitigation measures have been implemented. If this is not done, a range of negative impacts will have greater effect and positive impacts would not be enhanced.

Significance

This is a subjective indication of the importance of the unmitigated impact in terms of physical extent and time scale and indicates the level of mitigation required. This can be applied as a qualifier to both negative and beneficial impact.

Low: a short duration, site-specific impact of low intensity would be of low severity
Medium: a medium-term impact or site-specific impact and high intensity may have a medium severity in the unmitigated state
High: long-term or permanent impacts, those with regional influence or high intensity would create highly significant impacts

An impact of "low significance" will have only a limited effect on the environment, whereas an impact of "high significance" will have a major impact on the environment.

Sensitive or vulnerable environments or features as well as secondary and cumulative impacts were also taken into account during evaluation of impacts. Concerns and issues from stakeholders were addressed as potential issues. Impacts that may arise during the different stages of the project lifecycle are addressed below in this section. Mitigation seeks to find better ways of undertaking activities to minimize or eliminate negative impacts and enhance and maximize positive impacts on people and the environment.

Mitigation measures are described in terms of an impact's nature, extent, duration, intensity, and significance. Timing is referred to according to the project phases (operation and closure) and duration is referred to as short, medium, and long term. Permanent mitigation measures are not considered. The management measures are based on the finding of the EIA along with existing legislation and available technologies. It is the responsibility of management to become aware of any new environmental legislation and put in place any new statutory mitigation measures. The following section discusses each impact identified in the EIA under the respective titles, Construction and Operation phases, and makes recommendations for its mitigation if it is a negative impact and enhancement in the case of a positive impact.

Mitigation Discussion

Mitigation seeks to find better ways of underrating activities to minimize or eliminate negative impacts and enhance and maximize positive impacts on people and the environment. Mitigation measures are described in terms of nature, extent, duration, intensity, and significance, Timing is referred to according to the project phase (Operation and Closure), and duration is referred to as short, medium, and long term. Permanent mitigation measures are not considered.

The management measures outlined below are based on the findings of the EIA along with existing legislation and available technologies. It is the responsibility of management to become aware of the new environmental legislation and put in place any new statutory mitigation measures. The following section discusses each impact identified in the EIA under the respective titles, Construction and Operation phases, and makes recommendations for its mitigation if it is negative impact and enhancement in the case of positive impact.

4.24   POTENTIAL IMPACTS DUE TO SITE PREPARATION, LANDFILL CONSTRUCTION, AND MITIGATION MEASURES

The physical environment of the area earmarked for the proposed waste disposal facility has a slope gradient of about 20 m in stream ward; a section of the proposed area is characterized by recuperating pioneer vegetation which suggests that at some stage the vegetation had been cleared off. At the construction phase, the project activities which are most likely to generate environmental aspects and mechanisms capable of having environmental impacts are, digging and preparing the site, digging trenches, construction of additional drainage, and preparing of the first cell. The

TABLE 4.5
Impact Analysis in the Construction Phase

Nature	Extent	Duration	Intensity	Significance	Significance after Mitigation
Negative	Site	Short-term	Medium	Medium	Low

construction period will last about 2 months. Table 4.5 presents the impact analysis of the construction phase.

MITIGATION DISCUSSION

During the preparation phase, the area should be cleared of trees and grass, and then dug down to about 2m. During the excavation work, different layers of topsoil and subsoil should be stripped and stockpiled in separate heaps for eventual use in the rehabilitation exercise. Stockpiling of the soil in different piles helps to ensure that valuable properties in the topsoil are not diluted. The disturbance to the soil that would result from the planned activity is not considered significant in terms of extent, structure, depth, and or dynamics. Impacts to soil will be minimal as the soils have already been disturbed. The impact of this activity is assessed to be of MEDIUM significance, mitigation measures are expected to be effective, and significance of impacts should be reduced to LOW.

4.25 POTENTIAL IMPACT ON TOPOGRAPHY AND MITIGATION MEASURES

4.25.1 EXCAVATION

Land disturbance is the most visible impact on site location. Disposal facility construction requires significant disruption of the project area. The excavation of overburden and storage of cover materials result in removal or covering of soil and vegetation and the modification of the topography of the area. This leads to the creation of visual scarring of the landscape.

The moisture capacity of soils around the site may be altered due to increased lateral flow down slope, into the open disposal site area. This can lead to drying up of soils around the site, an associated increase in localized sheet erosion and eventual desiccation of soils, leading to the development of a dust bowl. Disturbance of soil in the form of sterilization occurs mainly through the compaction due to the use of heavy machinery. There may be increased potential for erosion in the drainage lines in and around the site due to the increased runoff from the site. Table 4.6 presents the impact analysis.

Mitigation Discussion

During the excavation works, different layers of topsoil and subsoil are stripped and stockpiled in separate heaps for eventual use in the rehabilitation exercise.

TABLE 4.6
Impact Analysis on Topography

Nature	Extent	Duration	Intensity	Significance	Significance after Mitigation
Negative	Site	Long-term	Medium	High	Medium

Stockpiling the soil from different layers helps to ensure that the valuable properties in the layer of the topsoil are not diluted. Soil erosion should be prevented in watercourse and drainage lines created. Curbing of soil erosion along the fringes of the main road is important, as the sandy texture of the rather coarsely grained soils could erode the gullies.

Rehabilitation helps to reduce the impact of the change in topography. The cleared area must be rehabilitated for decommissioning of the site. This impact significance is assessed as HIGH but reduced to MEDIUM with mitigation.

4.25.2 SURFACE WATER FLOW AND POLLUTION

The site is on a gentle slope and water flows down slope to the stream though there are no major drainage lines in the site itself. Clearing of vegetation and increased area of hard surface as a result of the development could increase runoff into natural drainage systems leading to flooding on the site. During the operation phase of the site, this can impact off-site discharge as indicated in Table 4.7.

Mitigation Discussion

During the operation phase, there is a need to ensure that activities do not impact the off-site ephemeral drainage. Adequate water management should take place in order to minimize the impact of increased runoff on drainage systems. A contaminated water pond needs to be constructed at a position whereby runoff water can reach by gravity. Contaminated water will evaporate from the pond. In times of high rainfall, the pond could overflow, but, in this case, the potentially polluted water would be sufficiently diluted to be acceptable in the environment. Contaminated water drains to the pond are to be constructed.

Upslope drains have to be constructed to divert clean runoff: an intercepting drain will divert any runoff to the pond via the main collector drain down the flank of the terraces. Soil erosion should be prevented. Continuous maintenance operation of site

TABLE 4.7
Impact Analysis on Surface Water Pollution

Nature	Extent	Duration	Intensity	Significance	Significance after Mitigation
Negative	Local	Long-term	Medium	Medium	Low

drainage and monitoring of all the activities is necessary. The impact significance is assessed as MEDIUM but may be reduced to LOW with mitigation.

4.25.3 GROUNDWATER POLLUTION RISK

Pollution of soil, surface, and groundwater resources may result from accidental spillage or even indiscriminate disposal of fuel and lubricating oils during vehicle maintenance operation. Use of heavy machinery necessarily entails the need to lubricate and service the equipment, which includes draining and replacing oil. This can be done on- or off-site. It's best to be done off-site. The area comprises mostly of complexly deformed gneisses and granulite of the Zambezi Belt's central zone.

No maintenance of vehicles is permitted on the site except in a controlled area with suitable service design to reduce the risk of oil entering natural drainage systems. In case of an emergency, any oil replacements should be made in such a way that the used oil is collected and disposed of at a designated point, especially oil recycling agents. Accidental oil and petrol spills from parked vehicles to be cleared up by disposal site staff. The impact of this activity is assessed to be of MEDIUM significance, and significance of impacts should be reduced to LOW.

4.26 POTENTIAL IMPACTS ASSOCIATED WITH GEOLOGY, SOIL, AND MITIGATION MEASURES

4.26.1 POTENTIAL IMPACTS ASSOCIATED GEOLOGY AND SOIL

The construction of a disposal facility requires excavation of the ground on a geologically stable base stratum to ensure sustainable environmental protection for the environment in close proximity. The present localized geological structures surrounded by fractured strata may act as preferential drainage paths for liquid contaminants.

The ease of excavation and potential drifting of the leachate from the disposal site into the underground are determined by the geology of the area. The greatest impact on the geology and soil associated with the construction of the disposal facility is groundwater pollution. This impact depends on the soil permeability and fractured strata. The alluvium exhibits a measured permeability of approximately 1.0×10^{-7} cm/s, and is classified as being very slightly permeable. This translates to an inferred percolation rate of approximately 0.00864 cm/day (approximately 0.25 cm/month).

The permeability rate depends on the soil texture overlying the strata. The soil permeability is high where the soil has high sand particle content and less in areas with high fines and fewer sand particles. The higher the fines content, the higher the water retention capacity, hence low permeability rate. The permeability potential of the soil is expected to rise during the construction process, as soil is exposed to evaporation and disturbance, resulting in soil loosening. Compaction should be conducted on the site.

Fractured strata may act as preferential drainage paths for liquids contaminants. In addition, soil loosening may, in the long run, result in gradual soil settlement under waste weight causing surface submergence at certain points on the disposal site top. The sites may create ponding during rain periods and eventually result in

TABLE 4.8
Impact Analysis on Geology and Soils

Nature	Extent	Duration	Probability	Significance	Status
Negative	Local	Long-term	Likely	Moderate	Medium

leachate generation. The impact associated with an increase in soil and strata permeability is accelerated leachate drifting into groundwater.

The potential pollution due to leachate generation is a localized impact, which is site-specific to the disposal facility site. Since the in-situ soil will be compacted by a roller to an acceptable permeability rate and imported low permeability soil can be added if necessary, a low permeability medium can be created resulting in reduced pollution risk due to leachate generation. Without the appropriate mitigation measures during construction, this impact is anticipated to be of moderate significance at the disposal site as indicated in Table 4.8.

The weakly alluvium exhibits a measured permeability of approximately 1.0×10^{-7} cm/s, and as such classify as being very slightly permeable. This translates to an inferred percolation rate approximately 0.00864 cm/day (approximately 0.25 cm/ month) (Bredenhann, 2005).

The study area is deemed to be moderately well placed with regard to surface water resources, as the non-perennial stream occurring approximately 100m downstream of the proposed facility is not deemed an important surface drainage feature, but rather acts as a conduit for enhanced sheet wash after heavy precipitation events.

Recommendation
In order to minimize the potential of pollution by leachate, the following mitigation measures are recommended. As far as possible, the excavated ground should be avoided from drying, probably by not exposing the surface to evaporation for a long time. Excavation during the winter more would be more cost effective and convenient.

4.26.2 Soil Excavatability

It should be possible to excavate the site to depth of at least 30 mm by hand or light mechanical excavator without serious difficulty, provided the excavation sidewalls are sufficiently reinforced to prevent localized collapse.

4.26.3 Soil Permeability

The topsoil covering the study area is generally deemed very permeable, with an expected percolation rate of approximately 1 m per month. However, the formation of large cracks in the topsoil due to shrinkage of the sand-rich soil material during periods of low rainfall will significantly increase the permeability of the top soil in localized areas. It must be noted that the presence of gneisses indicates the weak seasonal formation of perched water tables at a relatively shallow areas within the topsoil.

4.27   POSITION WITH REGARD TO DOMESTIC WATER SUPPLY SOURCES

Boreholes currently utilized for the abstraction of groundwater for domestic water supply and stock watering purposes are located at a distance of approximately 1 km opposite side of the proposed facility. The groundwater is deemed ideal to good. The underlying weathered and fractured gneiss aquifer is thus deemed an important local groundwater source and should be protected against contamination as it may be utilized in the future as a source of water, necessitating the placement of boreholes closer to the proposed facility.

4.28   POSITION WITH REGARD TO SURFACE DRAINAGE FEATURES

The study area is deemed to be moderately well placed with regard to surface water resources, as the non-perennial stream occurring approximately 100 m downslope of the proposed facility is not deemed an important surface drainage features, but rather acts as a conduit for enhanced sheet wash after heavy precipitation events.

4.28.1   SITE TOPOGRAPHY

The site generally exhibits gentle slopes that may require the use of a small retaining wall incorporating surface and sub-surface drainage precautions to secure, upslope of the waste disposal facility.

4.28.2   SITE DRAINAGE

The site exhibits low surface run-off velocities after precipitation events, and as such may be prone to surface erosion only in those areas where the natural drainage paths have been disturbed or a concentration of surface water occurs. It must be noted that localized ponding of water at the surface may occur after heavy precipitation events, especially in those areas where the surface topography has been landscaped. An efficient surface drainage system will therefore be required especially upstream of the site, in order to channel storm water around the facility.

4.28.3   FOUNDATION CONSIDERATIONS

In light of the results of the study, it can be stated that the site exhibits good mechanical properties that do not require the implementation of specialized foundation design and construction methods for structures to be erected at the proposed facility. The soil material occurring to a depth of at least 1.0 m, 2.0 m, and 3.0 m was found to be of good and non-expansive and/or compressible quality, requiring normal foundations.

4.29   POTENTIAL IMPACT ON WATER SYSTEM AND MITIGATION

The study area is located about 4 km away from the Zambezi River, about 1000 km from its source. Therefore, there are very remote chances of interaction. This is Zimbabwe's best-known geographical feature and tourist attraction center. The flow of water over the falls varies according to the time of year and the rainfall in the Zambezi's upper catchment areas in western Zambia and Angola. In November and December, it may be only 19 992 m^3 per minute but toward the end of a normal rainy season it reaches $5 \times 10^5 \, m^3/min$. During the record-breaking floods of 1958, it reached a peak of $7 \times 10^5 \, m^3/min$. The falls take the form of a sheer-sided chasm, as little as 60 m wide in places, running almost at a right angle across the river width. The southern tip of the gorge creates a natural viewing platform from which the full extent of the falls can be seen at close quarters.

The site is about 100 m from a stream. The construction of the disposal facility close to the stream potentially impacts on the water resources through water pollution. The study area is deemed to be moderately well placed with regard to surface water resources, as the non-perennial stream occurring approximately 100 m downstream of the proposed facility is not deemed an important surface drainage feature, but rather acts as a conduit for enhanced sheet wash after heavy precipitation events (Gorgens et al., 1997). The results of this study reveal that this site exhibits geological characteristics that will require the implementation of specific design and/or precautionary measures to reduce the risk of ground or surface water pollution from the proposed waste disposal facility as indicated in Table 4.9.

4.29.1   POTENTIAL IMPACT ON GROUNDWATER

The site exhibits a slight risk that liquids moving laterally through the soil horizons may reach groundwater sources.

The site also exhibits a slight risk that liquids from the site may reach groundwater sources occurring at depths as indicated in Table 4.10.

TABLE 4.9
Impact Analysis on Surface Water Drainage

Nature	Exte\nt	Duration	Probability	Significance	Status
Negative	Local	Long-term	Likely	Moderate	High

TABLE 4.10
Impact Analysis on Groundwater Drainage

Nature	Extent	Duration	Probability	Significance	Status
Negative	Local	Long-term	Likely	Slight	High

4.29.2   RECOMMENDATIONS

In order to minimize the potential of pollution by leachate, the following mitigation measures are recommended. The site requires the implementation of precautionary measures and a regular monitoring program. However, these measures do not disqualify the site from being used for the proposed development.

4.30   POTENTIAL IMPACT ON VEGETATION AND ECOLOGY

Mitigation calls for protecting and restoring as much of the original condition on the development site as possible. The planned use of land area partially addresses the loss of habitat and biodiversity by creating an ecological buffer zone. Additional measures must be considered to further minimize negative impacts on the terrestrial ecology in the area: If necessary, identified trees should be clearly marked and protected.

Given that a small portion of land would be developed, selecting plant species for replanting is not essential, as the types of birds, butterflies, and other fauna that are in the adjacent areas will naturally re-inhabit the site upon closure of the disposal facility. An integral part of the landscape plan should also address means of protecting and monitoring the area during site clearing and construction phases to ensure that the ecological integrity of the adjacent area is maintained.

RECOMMENDATIONS

Because a large portion of the adjacent land will remain undisturbed, it is expected that these areas will act as feeders or providers of seedlings for rehabilitation, allowing the displaced avifauna to remain and/or return to the general vicinity, thus maintaining the existing biodiversity. The area has no ecologically valuable trees, therefore no special mitigation measures are required.

4.31   POTENTIAL IMPACTS ON AVIFAUNA

4.31.1   HABITAT DESTRUCTION

An immediate and most adverse environmental impact to the area will occur during the preparatory phase which calls for clearing of the site for the proposed development. The removal of trees and shrubs will reduce the existing cover, resulting in loss of natural habitat for flora and fauna particular to the area. The proposed development will have insignificant negative effects on the composition of the portion of the farm. There will be obviously reduced visits to the development area by birds. The loss of habitat and negative impacts on the local biodiversity are obvious adverse consequences of the proposed development, although insignificant. Literature reveals that the area for development is an open dry veld. It is not used for nesting by birds but a source of food to birds and small mammals. Based on such an understanding, the forest patches are predominantly the source of livelihood and nestling for the birds in the area.

4.31.2   Disturbance

The increased traffic to the area, use of heavy equipment during the clearing of the site, and transportation of building materials will create noise and raise dust, which could further disturb the habitat of the existing fauna, in particular the birds, but also the plants and insects they feed on. Dust and emissions from the construction vehicles and heavy machinery are inevitable both during the site clearing as well as during construction phases. Airborne pollution, in particular dust, resulting from clearing of the land and from exposed piles of building materials (e.g. sand, cement) may further stress the local flora and fauna and pose a health risk to construction workers and residents in the vicinity who suffer from asthma or other respiratory ailments.

4.31.3   Bird Poisoning

Disposal sites are known to impact on bird species which feed on food waste and drink polluted water (leachate) at disposal sites. Species most vulnerable to poisoning tend to feed on domestic waste and are predominantly vultures. The daily covering of waste will hide waste and also impede ponding at the site. This will stop the birds from visiting the site as there will be nothing to feed on. Daily practice of waste covering is part of the mitigation measures and will minimize the impact of the disposal site on birds in the area.

4.32   POTENTIAL IMPACT ON THE SOCIAL ENVIRONMENT AND MITIGATION MEASURES

4.32.1   Employment Opportunities

The project will provide long-term jobs and some short-term construction opportunities. All the communities have expressed the need for employment opportunities. This impact is positive and will help improve the standard of living to the employees from the communities as is indicated in Table 4.11.

4.32.2   Influx of Job Seekers and Impact on Population Growth

Construction activities can attract people with varying skill levels to the site of the construction. Because the scale and length of this project is small and only for a short period of time, it is not expected to attract anyone.

TABLE 4.11
Impact Analysis on Employment

Nature	Extent	Duration	Probability	Significance	Status
Positive	Sub-regional	Short to Long-term	Likely	Medium	High

TABLE 4.12

Impact Analysis on Disruption in Daily Living

Nature	Extent	Duration	Probability	Significance	Status
Negative	Local	Short to Long-term	Likely	Medium	Medium

4.32.3 DISRUPTION TO DAILY LIVING

The daily living style may be insignificantly affected upon, although a few people will temporarily reside at the site during construction as indicated in Table 4.12.

To arrest disease spread due to poor sanitary conditions, portable toilets are to be utilized on the project site, and these facilities are to be properly and regularly maintained.

4.32.4 IMPACTS ON LAND AND RESOURCE USE

By providing a proper waste disposal facility, a positive impact will result for the industry, commercial farming, and planned industrial and commercial development, as well as residential, educational, and recreational facilities within the area. The disposal facility is urgently required by the residents. The disposal site therefore will have a high positive impact on future residential developments.

4.32.5 FORMATION OF ATTITUDES

Often communities do not like construction developments in their area, which have adverse effects on their lives and environment. Waste disposal is one such type of development which can be perceived as having a high negative impact rather than positive impact. Consultations with communities reduce the chances of negative attitudes.

4.32.6 NOISE POLLUTION

Noise and vibration from equipment operation and activities on site will raise the ambient noise levels in the area. The noise may disturb humans and animals. Noise from traffic during operation could have an impact on site neighbors and the general public. Noise emissions should, as far as possible, be controlled and noise levels of less than 85dB(A) must be maintained (Light and Broderick, 1998). All activity in the waste disposal site should be carried out during normal working hours and days (Monday–Friday, between 7 am and 5 pm). Hearing protection must be made available.

4.33 POTENTIAL IMPACT ON INFRASTRUCTURE AND SERVICES

The establishment of the waste disposal facility has a high positive impact on the community infrastructure and services. No roads were identified for closure or rerouting. Waste disposal will be significantly improved to cater to the whole community in an economic, efficient, and environmentally friendly way.

4.33.1   IMPACT ON LOCAL ECONOMY AND REGIONAL BENEFITS

In terms of the wider region, this disposal facility will meet the predicted increase in demand for disposal space in the town and improves environmental sustainability. From an employment perspective, both direct and indirect job opportunities will be created as a result of the construction and maintenance of the proposed disposal site. The job creation through entrepreneurship may be influenced by structured waste management service to elongate disposal site life span. These activities, such as recycling, create noticeable positive impact on the economy of the area.

The construction phase of the proposed development will employ a number of people from the local community and about five people will be employed during the operational phase as well. The biggest single impact that the proposed activity will have on the regional area is that of providing environmental sustainability to the public, whilst allowing economic sustainability to residents.

4.33.2   HEALTH IMPACTS: IMPACT FROM DUST PARTICLES

Land disturbance is the most visible impact of waste disposal construction. But waste disposal facility construction requires the excavation of overburden and storage of cover material. As a result, dust particles are thrown and suspended in the air (pollution). Generally, the activity leads to modification of the topography of the area, thus creating a visual scar on the landscape.

The moisture capacity of soils around the site may be altered, due to increased lateral flow down slope into the open disposal site area. This can lead to drying up of soils around the site, an associated increase in localized sheet erosion and eventual desiccation of the soils, leading to the development of a dust bowl. During the operation of the disposal site, dust particle release into the environment may result due to heavy machinery movement to and on the site, daily waste covering exercise, and waste tipping. Dust particles can cause respiratory problems to the people working on the site as well as people leaving in the area close to the waste disposal site. The polluted air may exacerbate the health problems of the people leaving with tuberculosis, asthma, and other respiratory ailments.

Recommendations

A dust control plan will be implemented. This will include the use of a bowser to periodically spray water at the construction site to suppress the dust. The bowser will also be used during the facility operation when the daily cover material is put on the waste top. People working on the site will be provided with protective gear to protect them from dust particles.

4.34   POTENTIAL IMPACTS ON HERITAGE RESOURCES

The cultural site that will be affected by the waste disposal facility is the cemetery. It has an average uptake of 3 adult graves per month and an average uptake of 10 children's graves per month. Under the heritage assessment, management actions

TABLE 4.13
Impact Analysis Heritage Resources

Nature	Extent	Duration	Probability	Significance	Status
Negative	National	Permanent	Probable	No significance	Medium

and recommended mitigation, which will result in a reduction in the impact on the sites, will be classified as follows (see also Table 4.13):

A – No further action necessary
B – Mapping of the site and controlled sampling required
C – Preserve site, or extensive data collection and mapping required
D – Preserve site

Impact analysis

Heritage significance: No significance
Impact: Negative
Impact significance: High
Certainty: Probable
Duration: Permanent
Mitigation: A

The study concluded that there are no primary or secondary effects at all that are important to science or the general public.

RECOMMENDATIONS

Heritage authorities should be informed of this development. If during construction material of possibly historical value is unearthed, a report should be submitted to relevant authorities.

4.35 ENVIRONMENTAL ISSUES

The relationship between the proposed development and its biophysical and socio-economic environment has been considered, and it is clear that the nature of the site and the surrounding area will be insignificantly altered through the construction of the proposed disposal site. The mitigation measures will ensure limited impact on the environment and future environmentally acceptable use of the site at the end of the disposal site life span. The identified mitigation measures must be adhered to. This will result in the minimization of negative impacts on the environment, and evidence shows that there will be no foreseen risks or hazards to the biophysical environment. Most of the mitigation measures, such as preventing contamination of soil, dealing with storm water, and minimizing runoff off the site, will aid in limiting the biophysical impacts of the development.

With respect to the socioeconomic impact of development, the importance of the proposed development is high as it will provide the disposal space necessary for the existing and growing demand in the area. The waste regulations require improved disposal facilities, specifically in areas where planned residential area growth is currently taking place. All these developments will ultimately employ people with various skill levels. Failure to construct a new disposal facility will have a severe negative effect on the environment and create health hazards.

4.36   ASSUMPTIONS, LIMITATIONS AND GAPS IN KNOWLEDGE

4.36.1   ASSUMPTIONS AND LIMITATIONS TO THE STUDY

The assumptions and limitations on which the study approach has been based upon include:

4.36.1.1   ASSUMPTIONS

All information provided by I and APs to the environment team was correct and valid at the time it was provided. The following sections outline the gaps in knowledge in each aspect of assessment.

4.36.1.2   Limitation – Seismicity Information

The design of a landfill site usually has to consider potential seismic activity at the site since it covers a large area and holds tons of waste. This requires the selection of an earthquake design for the site in question. A method commonly used to determine the effects of the earthquake design on a particular site is to assume that the earthquake occurs on the closest known possibly active fault (selected on the basis of the geological studies previously conducted in the area). Attenuation tables are then used to estimate the magnitude of the earthquake forces reaching the site as a result of the earthquake design occurring on the selected fault.

4.37   ENVIRONMENT ASSESSMENTS

4.37.1   AVIFAUNA ASSESSMENT

No information deficits or limitations were experienced. The literature and physical appearance of the area for development revealed that the area is an important bird area.

4.37.2   VEGETATION ASSESSMENT

No limitations in information acquisition for vegetation were experienced.

4.37.3   HERITAGE ASSESSMENT

Heritage resources can be found in unexpected places, and surveys may not detect all the heritage resources in a given project area. While some remains may simply be missed during surveys (observation), others may be hidden below the surface of the earth and not obvious until development (such as construction of the facilities and access roads) commences.

4.38   OPINION ON AUTHORIZATION OF THE LANDFILL

This section discusses whether the activity should be authorized or not and conditions that should be met in respect of the authorization.

4.38.1   TOPOGRAPHY

The fundamental topography of the study area and, in particular, the preferred disposal site will remain relatively similar to its current profile. The proposed activity will include a leveled surface on which the disposal site will be constructed. The environmental management plan (EMP) will be carefully followed, and there will be no significant impacts on existing study area topography.

4.38.2   GEOLOGY

The results of this study reveal that this site exhibits geological characteristics that will require the implementation of specific design and/or precautionary measures to reduce the risk of ground or surface water pollution from the proposed waste disposal facility.
   These include:

   i. The presence of localized geological structures surrounded by fractured strata that may act as preferential drainage paths for liquids contaminants.
   ii. A relatively thick cover of very slightly permeable topsoil with an inferred percolation rate of approximately 1 m per month with the occurrence of occasional large cracks in the topsoil close to the surface, during the dry season, that allow the rapid infiltration of liquid pollutants.
   iii. The lack of a suitable groundwater level and quality monitoring point at the site itself.

Although these characteristics do not disqualify the site from being used for the proposed development, it will however require the implementation of precautionary measures and a regular monitoring program.

4.38.3   SOIL

The soils of the site will be impacted upon by the proposed development. The impact on the soil is considered to be low. Soils will be exposed during construction of the disposal site, when holes are excavated for fence foundations, and during bush clearing operations for fences and access roads. These access roads will be used not only during construction but also for maintenance during the life span of the disposal site.

The existing access road will be used for access during construction (and future maintenance) of the disposal site, thereby minimizing the disturbance to soil. Geology and soils may be impacted upon by construction activities, in particular where cut and fill is required (although these instances are not expected to be very frequent). All attempts should be made to preserve topsoil from areas that will be disturbed during construction for use later on in the rehabilitation process. Anti-erosion measures in the design of the access roads should be taken into account. Since the development of the site prior to settlement will be undertaken responsibly and in accordance with best practices and an EMP, the impact on the soils on the site will be within reasonable limits and can be mitigated and rehabilitated after decom-missioning of the facility.

Any soil which is contaminated by hazardous materials, including fuels such as diesel, should be removed and taken to an appropriate disposal site. It may be most advantageous to hard surface the entire site and filter any storm water which collects on site through an oil/water separator.

The disposal site is relatively flat; hence no portion of the site should become par-ticularly vulnerable to erosion or damage. However, topsoil should be removed from the site prior to development and construction, and this topsoil should be spread on other portions of the site or used for landscaping of the entrance or other publicly visible areas. Topsoil must be stockpiled according to stipulations contained in the project EMP.

The cumulative impact of the loss of topsoil is a risk over the whole site area, including the disposal site. The significance of the impact is expected to be medium. To mitigate this impact, best practices, particularly stockpiling top soil and using it to cover adjacent areas before landscaping, quality control, and adherence to an EMP during implementation need to be followed.

4.38.4   FLORA

The proposed location of the new waste disposal facility area is dominated by open veld and bush patches. This is an ecologically sensitive area where birds, animals, or plants were identified that would render the targeted location unacceptable for con-struction of a waste disposal facility. Some vegetation will be cleared in the devel-opment site construction of the facility and the access road. Studies show that the impact will be significant since some endangered species exist in the area such as elephants.

4.38.5   FAUNA

Habitat destruction will be limited to the disposal area where vegetation will be cleared. The vegetation in the area is not red data, so no special attention should be accorded.

4.38.6   AVIAN DIVERSITY AND POPULATIONS

Disposal sites are known to impact on bird species which feed on food waste and drink polluted water (leachate) at the site. The daily covering of the waste will reduce the chances of the vultures and stock birds accessing food and polluted water at the site.

4.38.7   HYDROLOGY

The proposed drains should be designed to accommodate a rainfall event that has a 50% chance of occurring in any one year. The proposed drains should incorporate upgraded natural drainage routes to avoid blockages and potential flooding. Sinkholes and depressions should not be backfilled or used as drainage outfalls but rather maintained as green spaces within the development. Source control techniques such as harvesting roof runoff, permeable pavements and infiltration devices are proven techniques in a complete, comprehensive, and sustainable drainage plan. Dealing with the water locally not only reduces the quantity that has to be managed at any one point, but also reduces the need for conveying the water off the site.

4.38.8   HERITAGE RESOURCES

No further studies/mitigations are recommended for the proposed project and no place of archaeological or historical significance will be impacted by the proposed project. However, should any chance archaeological or any other physical cultural resources be discovered subsurface, heritage authorities should be informed. From an archaeological and cultural heritage resources perspective, there are no objections to the proposed waste disposal site project

4.39   ENVIRONMENTAL IMPACT STATEMENT

4.39.1   AVIFAUNA ASSESSMENT

The site will pose a significant threat to the avifauna environment. This is because the area is an important bird and animal area. Therefore, there will be significant impact on the bird species.

4.39.2   FAUNA ASSESSMENT

The town has a remarkable diversity of animals. It is particularly rich in cattle, donkeys, baboons, wild goats, wild pigs, hyenas, hippos, rhinos, elephants, nuisance mosquitoes, and domesticated goats. The endangered animals, like elephants, are well inside the area in which the proposed waste disposal facility will be located. Therefore, there will be significant impact on the fauna. The future disposal site will be fenced and covered to prevent any interaction with animals. Furthermore, it is proposed trenches be dug around it to prevent access by elephants.

4.39.3   VEGETATION ASSESSMENT

The site will pose insignificant or no threat to the vegetation environment. This is because the area is not an important vegetation area.

4.39.4   HERITAGE ASSESSMENT

The assessment reflected that there are cemeteries that may be negatively impacted by the site to be developed. The heritage authorities will be asked to develop the graves going in the opposite direction of the proposed disposal site.

4.39.5   GEOHYDROLOGY ASSESSMENT

The results of the geohydrological assessment indicate that the site will require the implementation of specific design and precautionary measures to reduce the risk of ground or surface water pollution from the proposed waste disposal facility. On its northeastern side, the disposal site will have surface drainage and a storm water diversion drain, at times the bottom of the slope. Due to this, the natural flow of the water will be used to develop the drainage system, whereby possibly polluted water will be separated from unpolluted water.

A leachate-detection system must be implemented to make sure that leachate is detected as soon as it might appear and diverted to the evaporation ponds. The professional operation of a waste disposal site does also require a monitoring system for surface and groundwater, which will indicate possible pollution. Therefore, water sampling points and monitoring boreholes will be installed. It is recommended to establish one borehole at the highest point of the property and another at the bottom of the future waste disposal facility.

4.39.6   GEOTECHNICAL ASSESSMENT

The greatest impact on the geology and soil associated with the construction of the disposal facility is groundwater pollution. This impact depends on the soil permeability and fractured strata. The alluvium exhibits a measured permeability of approximately 1.0×10^{-7} cm/s, and as such is classified as being very slightly permeable (Love et al., 2005). This translates to an inferred percolation rate of approximately 0.00864 cm/day (approximately 0.25 cm/month), which is lower than 3.7×10^{-7} cm/s, as stipulated by the minimum regulation requirements. From this perspective the chances of pollution are regarded as within natural control.

4.39.7   RISK ASSESSMENT

The assessment revealed that no significant negative impacts can result due to the operation of the facility. The site will have structures to monitor the risk associated with the facility operation.

4.39.8   SOCIOECONOMIC ASSESSMENT

The socioeconomic assessment revealed that the disposal site is among the high priority required facilities. This is shown by their high expectations and no negative comments were received with regard to the facility construction.

4.39.9 Visual Aesthetic Assessment

Assessments revealed that the site will not have an abstraction effect on the adjacent infrastructure. Its operation or existence will have no impact on the visual aspect of adjacent buildings or infrastructure.

4.40 CLOSURE PLAN FOR EXISTING DUMPSITE

As part of the closure, the project will try to minimize contamination caused by wind, leachates, or surface runoff. The project considers the use of biogas release vents. The following activities will be carried out for closure of the dumpsite:

First, preliminary planning will be carried out to define actions needed to prepare required closure plans, and the specification of engineering procedures for the project tasks. Later on, actions needed to schedule closure activities will be carried out, and regulatory agencies and users will be notified about the closure. During the closure phase, access to the site will be limited and signs will be put up to notify of the closure. Subsequently, light materials that are spread out throughout the site will be collected and taken to the dumpsite's main area to be covered with compacted material. The site must be leveled and furnished with an adequate slope for rainfall drainage (such as a 3% slope) (DWAF, 1998). Once the site has been leveled, wells for biogas venting will be installed. Finally, treatment must be provided to areas that have been affected by substances that have infiltrated the ground, such is the case with leachates. Within a period of three months after the site's closure, actions will be developed to finish drainage tasks such as those for the gas and leachate control and monitoring systems, and to install the cover and reforest the site.

Once the site has been closed, long-term maintenance will have to be implemented, especially in biogas and leachate monitoring and control systems, since these require continuous attention to ensure proper functioning. Some of the most important maintenance actions include cleaning leachate conveyance lines, cleaning storage lagoons, and inspecting existing pumps. The closure plan for the dumpsite will include:

  i. Laying of cover layer over the compacted waste followed by developing landscape on it
 ii. Landfill gas collection, venting and flaring system
iii. Leachate collection system
 iv. Surface water drainage for the storm water
  v. Sheet piling on the water ward side to prevent leachate from entering the sewage ponds
 vi. Construction of bunds, access roads and compound walls on the landward side of site

4.41 CREATION OF A RECYCLING PLANT

Recycling is a resource recovery practice that refers to the collection and reuse of waste materials such as empty beverage containers. The materials from which the items are made can be reprocessed into new products. Material for recycling may be collected separately from general waste using dedicated bins and collection vehicles are sorted directly from mixed waste streams. Recyclable material constitutes about 60% of the solid waste generated (Masocha, 2004, 2006).

The most common type of waste that can be recycled in The Town include aluminum such as beverage cans, copper such as wire, steel food and aerosol cans, old steel furnishings or equipment, polyethylene and PET bottles, glass bottles and jars, paperboard cartons, newspapers, magazines and light paper, and corrugated fiberboard boxes. Various types of plastics (PVC, LDPE, PP, and PS) are also recyclable. These items are usually composed of a single type of material, making them relatively easy to recycle into new products. PALENO is involved in collecting all the cans and plastics in the town, whereby these are sent to a recycling company in South Africa called Collect A Can. This means that no aluminum-based waste and plastics will find their way to the landfill.

Furthermore, Eco-composting is being implemented at the hotel. In Eco-composting, earthworms are used to biodegrade any organic material inclusive of food wastes into a bio fertilizer. This bio fertilizer is being used in the gardens at the hotel. It is therefore recommended that all the hotels, lodges, and households in the town adopt this strategy such that all the biodegradable waste will not go to the proposed landfill. These recycling strategies if adopted will increase the lifespan of the landfill.

4.42 ENVIRONMENTAL MANAGEMENT PLAN FOR PROPOSED LANDFILL

This is to guide planners, designers, operators, and regulators of the engineered landfill facility in the town. The document is organized in sections based on the sequences taken when developing the landfill facility. Each section provides technical information followed by the associated regulatory requirements.

These guidelines promote effectiveness and efficiency of municipal solid waste management, thereby reducing the overall cost of planning, design and operations and maintenance of the landfill facility while ensuring the protection of public health and the environment. These guidelines focus on objectives and principles rather than numerical limits. The latter are presented as recommended guides to summarize the available and current literature.

Landfilling is a method of disposing of solid waste on land in a manner that protects human health and the environment. Applying engineering principles, solid waste is confined to the smallest practical area, reduced to the smallest practical volume, and covered routinely with a cost-effective layer of earth. There are adverse health, safety, and environmental risks from open burning as a method of waste control. Open burning of hazardous wastes will release toxic substances into the atmosphere, potentially causing immediate health and environmental effects. This may adversely affect fire-fighting efforts. These substances may also harm the local ecosystem.

Burning can also spread quickly beyond the initial area, becoming a much larger problem. Pressurized vessels, such as aerosol cans and propane tanks, are an

explosion hazard and can become grenade-like projectiles. Open burning of municipal solid waste is not acceptable, except for clean wood and paper. When burning
these specific materials, the wastes should be moved to an area separate from the
working landfill. Permitting for burning is required from EMA. Burning should only
be done on days with very light wind or no wind.

4.43   OBJECTIVES OF SOLID WASTE MANAGEMENT

The community is recommended to adopt the 3R's of solid waste management:
reduce, reuse, and recycle. The objective of these activities is to divert as much waste
from landfill as is appropriate to the opportunities that exist. To meet this objective,
four major considerations must be addressed: public health and safety, environmental protection, costs, and aesthetics.

4.43.1   PUBLIC HEALTH AND SAFETY

Public health impacts may arise at all stages of solid waste management from collection to transport to disposal. The main concerns are (i) communicable diseases
transmitted from human fecal wastes disposed of via honey bags; (ii) uncovered
wastes promoting infestations of disease vectors (bacteria, insects, and rodents); and
(iii) the release of carcinogens and respiratory irritants from the incomplete combustion of open burning.

4.43.1.1   The Public Health Act

Regulations (Public Health Act Chapter 15:09) require that adequate solid waste facilities be provided and maintained so that there are no odors and no breeding of flies.

Regulations stipulate garbage containers must be provided and emptied regularly
and facilities must be situated at least:

  i. 90 meters from public roads, railways, right-of-ways, and cemeteries
 ii. 450 meters from housing
iii. A distance from water sources that ensures the protection of drinking water

Enforcement of the Public Health Act is through the Chief Medical Health Officer
and the appointed Medical Health Officers and Health Officers. Contravention of the
Act and its regulations by landfill operators could result in an order to comply, which
if refused may render the operator liable on conviction to a fine or imprisonment

4.43.2   ENVIRONMENTAL PROTECTION

Proper siting, design, and maintenance and operations of modified landfill facilities are fundamental in minimizing the environmental impacts associated with solid
waste disposal. Particular to the town are potential environmental impacts such as:

  i. Surface water and groundwater contamination; and
 ii. Improperly stored hazardous wastes.

The Environmental Protection Services (EPS), Department of Resources, Wildlife and Economic. With respect to solid waste facilities, the Environmental Protection Act is mainly concerned with hazardous wastes. EPS guidelines are available for specific substances such as waste solvents, antifreeze, asbestos, lead, lead-based paint, other paint and batteries. Hazardous waste receivers must be registered with EPS and follow the guidelines set out in the Guideline for the General management of the hazardous waste

If an infraction is detected by an inspector, the solid waste operator may be issued an order to stop the discharge of a contaminant by a certain date or to repair, remedy any injury or damage to the environment. Contravention of the Act by any person causing or contributing to the discharge, or the owner of the contaminant, may be found guilty and liable to a fine.

4.43.3   Aesthetics

The aesthetics of landfills, namely foul odors and unsightly facilities, are a concern for the public. Solid waste facilities should be sited far enough away from a community such that odors are not regularly detected and the site is not visible by the residents. In fact, the site is downwind of prevailing winds.

4.43.4   Regulatory Requirements

The regulatory requirements associated with municipal solid waste management are given in Table 4.14.

4.43.5   Planning Solid Waste Facilities

Planning modified landfills involved an understanding of the current as well as future requirements of communities, and applying engineering principles to design adequate solid waste facilities. Such planning must have considered the physical characteristics of solid wastes, siting considerations, surface and groundwater impacts, and projected population growth.

4.43.6   Physical Characteristics of Solid Wastes

For planning purposes, waste volumes, densities, compaction rates and composition of solid waste were determined for the community.

4.43.7   Density

There is a wide range of municipal solid waste densities quoted in the literature. For the town, a density of 0.099 tons per cubic meter for non-compact waste is acceptable. Other densities may be acceptable, but should be justified.

4.43.8   Industrial and Commercial Wastes

The management of industrial and commercial wastes is the responsibility of the waste generator. It is often disposed of using private facilities, but to the same landfill

TABLE 4.14

General Regulatory Requirements Associated with Solid Waste Management

Consideration	Acts, Regulations and Guidelines	Authority	Implication
Water Pollution Surface water and groundwater contamination. Improperly stored hazardous wastes.	Land, water and the environment act	All Land and Water Boards.	Water License. Enforcement of Water License. For Acts and Regulations: Order to comply. If order is refused, liable to fine or imprisonment. For guidelines: Recommendations Given
Air Pollution Surface water and groundwater contamination. Gas emissions from waste decomposition. Emissions from open burning; and improperly stored hazardous wastes	Environmental Protection Act. Guideline for the General Management of Hazardous Waste	Chief Environmental Protection Officer. Inspector.	For Acts and Regulations: Order to comply. If order is refused, liable to fine or imprisonment For guidelines: Recommendations given
Fire Open burning at site Wildfires	Fire Protection Act and Regulations	Office of the Fire Marshal.	For Acts and Regulations: Order to comply. If order is refused, liable to fine or imprisonment For guidelines: Recommendations given
Diseases Communicable diseases transmitted from human fecal wastes Uncovered wastes promoting infestations of disease vectors Carcinogens and respiratory irritants from open burning.	Public Health Act General Sanitation Regulations	Health and Social Services: Chief Medical Health Officer. Medical Health Officer. Health Officer	Order to comply. If order is refused, liable to fine or imprisonment
Safety Public safety from hazard to aircraft Establishing Guidelines for the Separation of Solid Waste	Commissioner's Lands Act Air Regulations and	Transport Zimbabwe. Municipal and Community Affairs Department of Transportation	Siting recommendation:

facility and is covered under separate guidelines and/or Water License requirements. The landfill should not accept industrial and commercial waste unless it conforms to the requirements and guidelines on discharge of industrial waste

4.43.9   HAZARDOUS AND BULKY WASTES

Household hazardous wastes may be an issue. The communities do not have the capacity or expertise to undertake the management of household hazardous wastes. Further, given the nature of many communities, there may be insufficient volume of household hazardous wastes to warrant an aggressive diversion program. However, the town will undertake an inventory of wastes before embarking on a program like that.

4.43.10   COMPACTION RATES

The recommended compaction rate for a modified landfill is 3:1 this rate varies widely but is the minimum expected for compaction when following recommended operations practices.

4.43.11   PLANNING HORIZON

This landfill was planned based on a 25-year planning horizon.

4.43.12   COLLECTION FREQUENCY

The preferred collection frequency for municipal solid waste (MSW) is once every week. Institutional and/or commercial collection frequency associated with MSW should be determined site-specifically.

4.43.13   COLLECTION PRACTICES

Cost efficiencies can be realized through more efficient routing and municipal pick-ups. Despite this variability, the following collection principles can be identified as having broad applicability throughout the town. These mechanisms include:

    i. The design of an effective waste collection system must consider the size of community proximity to neighboring communities, and proximity to landfill

    ii. The collection system may involve direct haul of waste from residences to landfill, or may include a transfer station where a central landfill is appropriate

    iii. Physical waste collection techniques will range from the use of small manual-load vehicles, to semi-automated or automated vehicles capable of handling both residential and commercial wastes

Each waste collection system must meet technical and financial requirements as well as public preferences and priorities. Convenience to users and level-of-service issues typically play a large part in the selection of the preferred system, and these aspects of waste collection cannot be meaningfully generalized. Technical requirements are susceptible to local geographic conditions (e.g. presence of year-round access), however the following general principles may be used for guidance:

i. Waste collection equipment should be selected according to the length of waste haul, frequency of collection, and the types and quantities of waste to be collected

ii. In communities where each residence uses an individual garbage can, collection service will be most efficiently delivered by 1 ton's compactor-type vehicles

iii. In communities where it is feasible for individual bins to service several residences, collection service may be delivered by 3 tons' side loader type vehicles. In this case, 1.15 cubic meter bins would typically be shared between 2, 3 or 4 houses. Operating efficiencies can be achieved in this system, since in addition to being used in the residential sector, the 1.15 cubic meter bins are large enough to be used by many commercial outlets (stores, offices) and consequently a single vehicle can be used to collect waste from both residential and commercial collection points

iv. Small communities with less than 1,000 residences will typically be most efficiently serviced by simple bin-style transfer stations, in which the bins are coated to prevent freezing of waste onto the container under winter conditions. Larger communities may benefit from more sophisticated compactor-style transfer stations, in which mechanical compaction is used to reduce the volume of waste prior to hauling for final disposal.

In general, cost efficiencies will be maximized when the following collection fundamentals can be combined:

i. Efficient routing
ii. Combined residential and commercial collection; and
iii. Optimized use of transfer facilities

4.43.14   Monitoring

4.43.14.1   Groundwater

To determine whether groundwater monitoring is warranted at the site, the following table should be used as a guide. The intent of this approach is to identify conditions that could reasonably be expected to represent risks to groundwater. The table is a guide only and local circumstances and professional discretion will dictate the final decision. Disputes or uncertainties should be resolved by a qualified Groundwater Scientist.

4.43.14.2   Surface Water

All land and water boards will require routine surface water monitoring program. At minimum three sampling stations will be required: upstream; immediately downstream; and at a receiving body.

4.43.15   DESIGNING SOLID WASTE FACILITIES

General

To minimize public health and environmental hazards a solid waste landfill is used for land disposal of refuse. This is done by periodically spreading the refuse into thin layers, compacting the refuse by driving over it a few times, and then applying a granular cover material. A modified landfill increases the interval between covering operations to once a month or even once a year. Design and operation are intended to ensure that the final landfill form is domed to promote the rapid runoff of surface water.

Design Life

The landfill is designed for a minimum 25-year design life.

Fencing

The installation and maintenance of fencing are recommended at the landfill facility for the following reasons.

i. To control or limit access to the landfill site by community residents
ii. To prevent scavenging animals from causing a nuisance and risking the safety of workers and residents
iii. To control the spreading of blowing garbage

Fencing may be portable or permanent and may be woven or chain linked. Electric fencing has proven to be an effective deterrent against bears. Wooden fences are not recommended, as they can be a fire hazard. Gating the entrance and providing times for the public to enter the site are also recommended.

Signage

Water licenses require that signs be posted in the area to advise the public that the site is being used for the disposal of solid waste. All accesses should be posted. Additionally, signs are appropriate to advise the public of the use of recycling and take-it-or-leave-it facilities, hours of operation, emergency numbers, and the like.

4.43.16   WATER

Flowing water should be prevented from entering the site. Cut-off berms, swales, and trenching are effective diversion methods. Water should not pool on site, rather drain quickly without causing erosion.

4.43.16.1   Monitoring Well Design

If required, a groundwater monitoring program must have a sufficient number of monitoring wells, installed at appropriate locations and depths, to yield water samples that:

i. Represent the background conditions of the site (usually hydrologically up gradient from the solid waste facility)

    ii. Represent the quality of groundwater passing through the site
   iii. Detect any contamination of the uppermost aquifer

The number, spacing, and depths of monitoring wells must be based upon the site-specific characterization of aquifer thickness, groundwater flow rate and direction (including seasonal and temporal fluctuations) Further, the saturated and unsaturated geologic units and fill materials overlaying the uppermost aquifer and lower aquifer characteristics must be taken into account including: thickness, stratigraphy, lithology, hydraulic conductivity, porosity, and effective porosity. Groundwater monitoring wells should be designed to best ensure they detect contamination from the solid waste site, that is, they should not be located at such a distance from the site as to avoid likely contamination. The system must be designed and submitted by a qualified groundwater scientist.

4.43.16.2   Hazardous Waste Storage Facilities
Hazardous waste storage areas should be located within a fenced area, separately fenced if possible, away from innocent public access. Based on the inventory of wastes, sufficient and separate storage should be provided

4.44   LANDFILL OPERATION AND MAINTENANCE

4.44.1   LANDFILL GENERAL CONSIDERATIONS

The cost-effective operation of a modified landfill is the basis for design. Operations are recommended as follows:

    i. Compaction rates of 3:1 or better are achieved by working a bulldozer or other appropriate heavy equipment over the waste 3 to 5 times
   ii. Compaction of wastes is undertaken once per week or in combination with collection frequency. Generally, the wastes should be worked and compacted as they are dumped
  iii. Operations should be undertaken to minimize close-out requirements
  iv. Cover material is generalized as 100 mm between cells, 300 mm on the surface of cells, 600 mm as part of close out

4.44.2   LANDFILL OPERATIONS

There are three main methods of operating a modified landfill, the area method, the trench method, and the depression method.

4.44.2.1   Area method
In the area method, waste is emptied out of collection vehicles at the bottom of a short berm. The berm should be 2 m high. Wastes are worked and compacted against the berm, when the compacted garbage is 3 m wide; the compacted wastes are covered with 300 mm of granular material. Dry, sandy material is preferred where available. The process is repeated until the landfill is full.

At this point the site can be closed out with a 600 mm cover, domed to promote runoff of water. Alternatively, a mound can be implemented by packing 300 mm of granular material on the surface prior to beginning the second layer. The environmental issues related with operating the landfill are indicated in Table 4.15. Each environmental issue identified under each assessment class aspect during operation of the landfill is characterized in Table 4.15.

4.45   COMMUNITY WORKS MANAGEMENT SYSTEM/MAINTENANCE

The community works management system/maintenance is a task-based maintenance management system to be developed. The system is made of several parts, each contributing to the overall running of the system. The parts include:

   i. An inventory of assets to be maintained
   ii. Quality standards to which assets are to be maintained
   iii. Maintenance procedures and production levels
   iv. A work order system to authorize work
   v. A maintenance schedule
   vi. Stock control
   vii. A method to collect data and report results
   viii. A method to develop annual budgets and work programs

4.46   MONITORING PROCEDURES

4.46.1   WEIGHT/VOLUME

The annual weight and/or volume of waste disposed should be determined and recorded. Using weigh scales is best practice; however, volumes can be estimated from truck box measurements.

4.46.2   HAZARDOUS WASTE STORAGE

All hazardous wastes entering and leaving the site should be identified, inventoried, and logged. No hazardous wastes are to be disposed at modified landfills.

4.46.3   REGULATORY REQUIREMENTS

As part of due diligence, and compliance, the operation will require the preparation of an Operation and Maintenance Manual for the Facility. The stated purpose of the manual is to assist community staff in the proper operation and maintenance of their waste disposal facility. It must include:

   i. A description of how facilities are operated and maintained
   ii. How often these tasks are performed
   iii. Who is responsible for their completion?

TABLE 4.15

Environmental Considerations during Operation of the Proposed Landfill

Environmental Aspect	Nature of Impact	Extent of Impact	Duration of Impact	Probability of Occurrence	Degree of Reversal	Degree of Irreplaceable Loss	Mitigation Measures	Level of Significance after Mitigation
Dust pollution	Negative	Local	Long term	High	Medium	Low	Use of Water for suppressing	Low
Groundwater pollution	Negative	Sub-regional	Long term	High	Medium	Medium	Use of liner	Low
Waste scattering	Negative	Local	Long term	High	Medium	Medium	Daily covering layer	Low
Soil contamination	Negative	Local	Long term	High	Medium	Medium	Use of liner	Low
Storm water	Negative	Local	Short term	Low	Medium	Low	Use of barrier	Low
Run off the site	Negative	Local	Short term	Low	Medium	Low	Use of collection system	Low

The manual must also demonstrate to the Water Board that the community is capable of operating and maintaining their waste sites. Inspectors will use the community's manual as part of their inspection procedure to ensure that the stated procedures are being undertaken. The manual is to be developed according to the requirements of the Water Board. For solid waste sites, it shall include but not be limited to the following:

    i. Location of facilities and proximity to receiving waters
    ii. Frequency of inspection of dams, dykes, and drainage courses
    iii. Controlling effluent discharge quality
    iv. Runoff and drainage control within and around the facility, and restoration of erosion
    v. Treatment of contaminated drainage
    vi. Prevention of windblown debris
    vii. Managing hazardous waste
    viii. Segregation of domestic, metal and recyclable waste materials
    ix. Method and frequency of site maintenance
    x. Alternatives designed to prevent burning of MSW

4.46.4 Due Diligence

Regulatory compliance requires due diligence. Due diligence may be as defined as:

    i. Establishing a proper system to prevent contravention of regulatory standards
    ii. Taking all reasonable steps to ensure effective operation of that system

As part of the due diligence program, an internal monitoring and reporting program must be put in place. This program actively promotes the interrelationship between staff and management so that information, resources, and finances can be directed effectively. Documentation is fundamental to the program. Some of the information requirements include:

    i. The Environmental Policy
    ii. Roles, responsibilities, and authorities
    iii. Significant environmental aspects
    iv. Legal requirements
    v. Training
    vi. Communication within the organization
    vii. Communication with regulatory authorities
    viii. Emergency response
    ix. Monitoring and measurement
    x. Audits
    xi. Records management

The Environmental Management Plan (EMP) for the proposed landfill is in Table 4.16. Environmental Management Plan that complies with The Environmental

Management Act (Chapter 20:27) was enacted in March 2003. The Act provided the legal backing that was lacking and Sections 97–108 of the Act deal with EIA provisions. Section 98 of the Environmental Management Act, Chapter 20: 27 provides that the project developer should engage an independent consultant to undertake the EIA. Statutory Instrument 6 of 2007 also provides for the implementation of EIA.

4.47 CLOSURE AND POST CLOSURE PLAN OF THE PROPOSED NEW LANDFILL

4.47.1 LANDFILL CLOSURE

Once a landfill has reached capacity, final closure must be completed in a manner that ensures the long-term protection of the environment. Site closure requirements include:

All land and water boards require notification of a pending closure. Generally, a plan must be submitted for approval at least six months prior to closure that includes the following information:

  i. Future land use
  ii. Leachate prevention and monitoring
  iii. An implementation schedule
  iv. Mapping which shows all disturbed areas, borrow material areas, and site facilities
  v. Consideration of altered drainage patterns
  vi. Type and source of cover materials
  vii. Hazardous wastes including waste oil
  viii. Contaminated site remediation

4.47.2 FUTURE LAND USE

Any future use of the site should be passive to reduce problems that may result from the stored waste. Recommended uses include:

  i. Waste transfer station or related storage area
  ii. Bulky waste storage
  iii. Passive recreation
  iv. Open area

4.47.3 INFRASTRUCTURE AND EQUIPMENT REMOVAL

Materials stored for reuse may be landfilled, including tires, wood, metal, and the like:

  i. Waste oil and other liquids should be identified and removed
  ii. Batteries and other hazardous materials should be identified and removed
  iii. Any buildings on site should be decommissioned and removed

iv. Fences should be removed and reused if possible, otherwise landfilled

v. Bulky waste should be removed

4.47.3.1   Grading and Capping

The site should be capped with 600 mm of material and graded to positive drainage. Where readily available, clay material is preferred. The site should be seeded to stabilize the soil and prevent erosion.

4.47.3.2   Survey

A final survey should be undertaken to mark designated areas, locate monitoring wells, map and document the extent of the site. The survey should be tied to a permanent benchmark if available.

4.47.3.3   Registration

The site can be identified as a Solid Waste Management Facility on the land title documents. A sign should advise the public that the site is closed and should indicate alternative facilities.

4.48   LANDFILL POST CLOSURE

Long term care of the decommissioned landfill is important so that the impacts to the surrounding environment are minimized. For every site, cover material should be allowed to settle and re-graded as necessary. Any cave-ins should be filled in to prevent standing water. Vegetation should be monitored to ensure that it continues to grow. There should be on-going maintenance of drainage pathways and the like.

4.48.1   INSPECTIONS

A post-closure inspection checklist should be filed with the appropriate regulatory body. The checklist should include:

i. Inspection frequency

ii. Items to be inspected

4.48.2   POST CLOSURE MONITORING

Operational monitoring shall be continued into the post-closure period until one or more of the following conditions apply:

i. It can be demonstrated that the site is no longer releasing contaminants

ii. It can be demonstrated that the site has reached an equilibrium state in which contaminant release poses no unacceptable risk to the environment

Proponents shall submit a report to the appropriate regulatory body which justifies the cessation of monitoring.

TABLE 4.16
Environmental Management Plan for the Landfill

ENVIRONMENTAL MANAGEMENT AND MONITORING PLAN

Impact Requiring Mitigation	Management Action	Responsibility	Monitoring and Auditing
Disposal site operations cause dust impacts. Air quality – Trucks lead to emission of dust and fumes	**Target: To minimize dust impacts**	Consultant	Responsibility
	Planning Phase: Dust suppression measures to be included in operating plan.	Site Operator	Municipality
		Municipality	**Monitoring**
	Operating and Remediating Measures:	Municipality	**Action**
	On-site gravel roads to be wetted when	Municipality	Municipality to
	dusty. Complaints register to be kept.	Municipality	inspect the
	Establish a disposal site monitoring	Site Operator	condition of
	committee. Road signs on the road to		the road
	indicate that big trucks will be using the		Monitoring
	road. Maintenance of traffic warning signs		Committee to
	to be in position, particularly at the		assess the
	intersection with the main road. Speed limit		complaints
	warnings before the turn off to the site.		register
	Maintenance of access road and road on site		
	to be in trafficable conditions at all times,		
	including wet weather.		
	Site Closure and Aftercare		
	No action required.		
	Standards: Minimum requirements for waste disposal by disposal site.		
Disposal site operation Control of nuisance, litter, fires. Aesthetics	**Target: To minimize nuisances**	Site Operator	Responsibility
	Planning Phase	Site Operator	ZINWA
	Measure to control nuisance to be included in	Site Operator	Municipality
	the operating plan (litter, odors, fires)	Site Operator	**Monitoring**
	Operating Measures	Site Operator	**Action**
	Ensure proper sanitary disposal site practice	Site Operator	Inspections
	of spreading, compaction, and covering of	Site Operator	must be held
	waste. Daily covering of waste cells and	Municipality	annually, at a
	weekly covering of trenches. Litter	Site Operator	minimum
	containment within the site by means of	Site Operator	
	waste compaction and cover. Collection of		
	litter on and around the site on daily basis.		
	Prompt covering of putrescible waste.		
	Clearing and maintenance of 5m width		
	firebreak around the fence. No burning of		
	waste material on the premises of the waste		
	disposal site or in the vicinity thereof.		
	Complaints register to be kept. Checking of		
	empty waste vehicles and cleaning of all		
	loose litter before leaving the site.		
	Progressive rehabilitation of the site.		
	Scattered tree planting around the site		

(Continued)

TABLE 4.16 (CONTINUED)
Environmental Management Plan for the Landfill

ENVIRONMENTAL MANAGEMENT AND MONITORING PLAN

Impact Requiring Mitigation	Management Action	Responsibility	Monitoring and Auditing
	Site Closure and Aftercare: Maintenance of firebreaks. Maintenance of trees and vegetation Standards Operating plan Minimum Requirements for waste disposal by disposal site		
Record keeping	**Target: To prevent pollution of water resources** **Planning Phase:** No measures **Operating Measures:** Keeping records of all waste entering the site. Categorize waste by the number of loads, defined by mass and type. Record keeping on both daily and a cumulative basis. Site reporting structure and lines of authority to be clearly defined. **Site Closure and Aftercare:** No action **Standards:** Disposal operating plan Minimum requirements for waste disposal by site	Site Operator Site Operator Site Operator Site Operator	Responsibility ZINWA Municipality Monitoring Committee **Monitoring Action** Municipality and Monitoring Committee to regularly inspect waste records
Water quality monitoring and site drainage	**Target to prevent pollution of water resources** **Planning Phase** Indicate drainage or runoff in Design Indicate drainage for contaminated water in design. Indicate drainage for clean water in design	Consultant	Responsibility ZINWA Municipality
Waste acceptance	**Target: To prevent disposal of unauthorized waste on site** **Planning Phase** Classify the type of disposal sites	Consultant	Responsibility ZINWA Municipality **Monitoring Action** ZINWA and Municipality to regularly inspect drains

(Continued)

TABLE 4.16 (CONTINUED)
Environmental Management Plan for the Landfill

ENVIRONMENTAL MANAGEMENT AND MONITORING PLAN

Impact Requiring Mitigation	Management Action	Responsibility	Monitoring and Auditing
Security	**Target: To prevent entry of unauthorized personnel on site** To minimize the risk to the public **Planning Measures** Indicate fencing and firebreak requirements **Operating Measures** Maintenance of the security fence Keep facility locked after hours Prevent entrance of salvagers and unauthorized persons **Site and Aftercare** Maintenance of the security fence and keep the gate locked **Standards** Disposal site operating plan Minimum Requirements for waste disposal by disposal site	Consultant	**Responsibility** Municipality Monitoring Committee
Continued disposal of waste in the site destroy remaining natural environment Remediate areas are invaded by exotic plant species	**Target: Prevent further degradation of the environment** **Planning Phase:** Indicate rehabilitation plans. Rehabilitation plan to indicate that only indigenous grasses to be planted in areas cleared of waste. Approve the site of rehabilitation plan before implementation of remedial measures **Operation and Rehabilitation:** Hydro seed areas with indigenous grass species after clearing waste or landscaping. Remove all exotic invader species on a continuous basis especially before seed formation. **Site Closure and Aftercare:** Continues to remove any exotic invader species **Standards:** Disposal Site Operating and Remediation Plan	Consultant Consultant Site Operator Site Operator Site Operator Municipality	Responsibility Municipality Monitoring Committee **Monitoring Action** Inspection of vegetation on and of site.
Capping Maintenance	**Target: To avoid soil and water pollution** **Planning Measures:** Specify cover material requirements. Specify covering technique **Operating Measures**: Inspection and maintenance of rehabilitated areas. Filling in of settlement depression or cavities caused by fire	Consultant Consultant Site Operator Site Operator Site Operator Municipality	Responsibility Municipality Monitoring Committee **Monitoring Action**

(Continued)

TABLE 4.16 (CONTINUED)
Environmental Management Plan for the Landfill

ENVIRONMENTAL MANAGEMENT AND MONITORING PLAN

Impact Requiring Mitigation	Management Action	Responsibility	Monitoring and Auditing
	Site Maintenance and Aftercare: Inspections to monitor cover integrity, subsidence, and vegetation **Standards** Disposal Site Operation Plan		Inspection of cover integrity, subsidence, and vegetation
Disposal Site Auditing	**Target: To ensure maintenance of acceptable standards** **Planning Phase** Indicate frequency and type of audits to be carried out **Operating Measures** Audit site on regular basis **Site Closure and Aftercare** Ongoing inspection of the site **Standards** Disposal site operation plan Minimum requirements or waste disposal by disposal site	Consultant Site Operator Municipality	Responsibility ZINWA
Facilities on site	**Target: To ensure maintenance of facilities to acceptable standards** **Planning Phase** List of facilities to be maintained (road, drains, contaminated water pond, gatehouse, and fencing) **Operating Measures** Maintenance of all facilities on site including the access roads **Site Closure** On-going inspection of the site **Standards**: Disposal site Operation plan. Minimum requirements for waste disposal by site	Consultant Site Operator Municipality	Responsibility Monitoring Committee ZINWA Consultants Municipality **Monitoring Action** Inspect condition of facilities on site
Health and safety risks	**Target: To minimize the risk of illness or diseases to workers as a result of the site operation** Abiding by NOSA or OSHA requirements. All health and safety measures as stipulated in current legislation to be followed at all times	Consultant Site Operator Site Operator Site Operator Site Operator Municipality	Responsibility ZINWA Municipality Monitoring Committee

(Continued)

TABLE 4.16 (CONTINUED)
Environmental Management Plan for the Landfill

ENVIRONMENTAL MANAGEMENT AND MONITORING PLAN

Impact Requiring Mitigation	Management Action	Responsibility	Monitoring and Auditing
	Planning Phase		**Monitoring Action**
	Stipulation of health and safety requirements.		
	Operating Measures		Inspection of
	Provision of workers with appropriate protective clothes		health and safety
	Appropriate safety awareness signs to be prominently displayed on the site		standards on site
	Fire extinguishers to be provided on the site		
	Preparation of an emergency plan for any unforeseeable emergencies such as accidents/ outbreak of infectious diseases		
	Provision of clean portable water for use by personnel		
	No unauthorized personnel should be allowed on site		
	Site Closure and Aftercare		
	No access of the public to the site		
	Standards		
	Disposal site Operation Plan		
	Minimum requirements for waste disposal by site		
Employment opportunities and conflict management	**Target: To minimize public conflicts regarding the site**	Consultant	Responsibility
	To employ as many local workers as possible	Site Operator	Municipality
	Planning Phase	Municipality	Monitoring
	Specify employment strategy	Municipality	Committee
	Specify measures to minimize conflict between the communities and the municipality regarding the project	Municipality	**Monitoring action**
		Council	Assessment of
		Town Council	who works in
	Operating Measures	and	the site and
	Employment of locals for unskilled and semiskilled positions	Municipality	who is subcontracted
	Provision equitable employment opportunities for affected communities		
	Hiring of local contractors to work at the site		
	Establishment of disposal site Monitoring Committee		
	Liaison between the councillors and the Municipality		
	Site Closure and Aftercare		
	Liaison between municipality and community		
	Standards		
	Disposal Site Operating Plan		
	Minimum requirements for waste disposal by site		

4.48.3   REGULATORY REQUIREMENTS

All land and water boards will require routine reporting. Generally, the requirements will be outlined in a license. While the typical design outlay cannot be achieved in Zimbabwe due to limited resources, the report provides a typical scenario of disposal site operation. A landfill must be open and available every day. Customers are typically municipalities and construction/demolition companies, although residents may also use the landfill. Near the entrance of the site is a recycling center where vehicles can drop off recyclable materials (aluminum cans, glass bottles, newspapers, blend paper, and corrugated cardboard). This helps to reduce the amount of material in the landfill. Some of these materials are banned from landfills by law because they can be recycled. Along the site, there are drop-off stations for materials that are not wanted or legally banned by the landfill. A multi-material drop-off station is used for tires, motor oil, lead-acid batteries, and drywall. Some of these materials can be recycled.

In addition, there is a household hazardous waste drop-off station for chemicals (paints, pesticides, other chemicals) that are banned from the landfill. These chemicals are disposed of by private companies. Some paints can be recycled and some organic chemicals can be burned in incinerators or power plants. Other structures alongside the landfill are the borrowed area that supplies the soil for the landfill, the runoff collection pond, leachate collection ponds, and methane station. While this is an optimal scenario, the reality is quite different for Zimbabwean Communities/ Municipalities, where resources are limited.

4.49   LANDFILL TECHNICAL INFORMATION AND DESIGN

4.49.1   SITE DESIGN

The site design is a process which follows the site selection and investigation process. Finally, the disposal site design will be based on the outcome of the site investigation as well as the EIA process. One of the most important reasons for the disposal site design is to provide a cost effective, environmentally acceptable waste disposal facility. In case that the best available site, which was identified during the site selection process is not as good from the environmental or geo-hydrological point of view, the responsibility from the design would be to compensate for the shortcoming in the most appropriate way. It is distinguished into two different stages of design, which are conceptual and the technical design. Thus, this section will concentrate on the site layout and the technical design. The scope of work included the following:

  i. Liner design and leachate management plan
  ii. Sub-surface and surface drainage systems
  iii. Capping and re-habitation design
  iv. Permanent storm-water diversion and anti-erosion measures

Based on the site investigations and findings the new site was classified as a G: M: B + (South African Standards). The total area is about 100 000 m². The total available air space is about 50 000 m² × 3.12 m = 156 000 m³ and based on the calculations it is anticipated that the landfill will be operational for the next 25 years at least.

The recommended designs are as follows:

i. A lining system comprising of an average of 300 mm leachate collection layer, layers of compacted soil of low permeability to prevent leachate from migrating into the ground, a geo-synthetic clay liner (GCL) through which the landfill leachate should seep through and a 150mm base preparation layer
ii. Separation liner between the existing waste body and the new extension comprising of a 300 mm leachate collection layer, geo-synthetic clay liner, and a 200 mm slope preparation layer
iii. Sub-surface drainage network comprising of 110 mm diameter perforated (high density polyethylene) HDPE
iv. Leachate collector pipes at 25m centres that drain into 200 mm diameter perforated HDPE pipes then tie to the existing leachate collection system
v. Surface drainage system consisting of earth berms, chutes and toe drains to prevent surface water on top of the landfill from ponding

4.49.2   LANDFILL SITE LAYOUT

Some of the site infrastructure already exists, for example the road access, which is provided via the tarred road which connects with the Bulawayo Road. On the southern side the disposal site will have surface drainage and a storm water diversion drain, which is at the same time the bottom of the slope. Due to this the natural flow of the water will be used to develop the drainage system, whereby possible polluted water will be separated from unpolluted water. Leachate management is required and a leachate detecting system has to be implemented to make sure that leachate is detected as soon as it appears. Professional operation of a waste disposal site also requires a monitoring system for surface and groundwater, which will indicate possible pollution. Therefore, water sampling points and monitoring boreholes will be installed. It is recommended to establish one borehole at the top of the property which is the highest point and another borehole should be installed at the bottom of the future waste disposal facility. The site will be fenced; a gate and gate house will be installed as well. This will avoid unauthorized access to the site. The fencing will be 2m high. The gate will be locked after working hours. During evenings and at night security will be present. Administrative facilities will be implemented. Therefore, an office block will be built, which includes changing rooms for the staff working at the disposal site, toilets, and showers as well as one office for doing all the administrative work which is required at a waste disposal site. Currently there is no municipal piped water and electricity at the site. Either running water connection need to be provided to the area or water tanks for drinking purpose need to be established.

4.49.3   LANDFILL LAYOUT PLAN AND DEVELOPMENT PLAN

The new waste disposal facility will be developed in phases, whereby each phase consists of one cell. Within the first phase of the waste disposal development, phase one will be established. This is the most southern phase, close to the access road and on the bottom of the slope. The first cell which is going to be developed within the first phase will be the most southern cell as well. An entrance road will be developed. The access to the road will be only available via the security gate, since the entire facility will be fenced. A service road inside the premises surrounds the area and 3 m wide fire break will surround the entire landfill site area.

Just after entering the site, the collection vehicle will enter the first cell by passing the office and ablution facility (to be constructed). Thereafter the truck will pass the installed weighbridge and the office where all the information regarding the truck, the amount and type of loaded waste are registered and maintained.

The storm water ablution pond will be opposite the main office. The location of the pond was calculated (details in the facility design) and based on the slope of the area. The soil which will be excavated within the first cell, to provide airspace for disposal, will be heaped close so as to use it for daily cover. This means the first cell will be excavated and prepared for disposal. While the first cell is in operation the second cell will be excavated. The landfill layout plan and development plan will then include the following things.

- Infrastructure
- Site access and drainage
- Excavation and stockpiling of cover
- Screening berms and screening vegetation.
- Cell construction sequence
- Deposition sequence and phases

4.50   LANDFILL REHABILITATION PLAN AND END USE PLAN

The intended end-use for the site is currently envisaged as open space. The domed shape should enable runoff to drain freely off the disposal site without penetrating beyond the topsoil.

4.50.1   SURFACE DRAINAGE DESIGN

It is proposed that surface water on top of the landfill needs to be diverted down by a system of earth berms, chutes, and channels. As shown in Figure 8.0, the earth berms are oriented in such a way that enhances flow of water. Any excess water from the earth berm will be diverted down by the chutes to the toe drains. The upslope cut-off drains must divert clean storm water around the site into the natural drainage system. The chute systems will have energy dissipaters downstream to protect downstream areas from erosion by reducing the velocity of flow. All drains should be maintained ensuring that they are not blocked by silt or vegetation. The trenches will have to be cleaned regularly because even small deposits of silt reduce the capacity of drainage.

4.50.2   DESIGN FOR SUB-SURFACE DRAINAGE

To ensure that no sub-surface leachate accumulate, the leachate should gravitate to leachate collector pipes that drain to a series of leachate drainage networks which lead to the existing sewer system. Based on the design calculations, 110 mm diameter perforated HDPE pipes at 25 mm centers are recommended for sub-surface drainage. The leachate should gravitate into 200 mm diameter perforated HDPE pipes then tie to the existing leachate collection system as shown in the detailed drawings. The new leachate collection system is to connect to the existing sewer system with manholes whose average invert level is 2.9 mm. A provision of inspection manholes has been made and they are well spaced at 75 mm.

4.50.3   Monitoring System Design

The boreholes that will be located in the proximity of the proposed site will be used for monitoring the system design.

Leachate Monitoring and Treatment

One of the most important problems associated with the design, operation, and long-term care of landfills is managing leachate that is formed when water passes through the deposited waste. The leachate generated from municipal solid waste is a mixture of organic and inorganic, and dissolved and colloidal solids. It contains products of decomposition of organic materials and soluble ions which present a potential pollution problem for surface and groundwaters.

Leachate generation rates are primarily dependent on the amount of liquid the waste originally contained, a quantity of precipitation that enters the landfill through the cover or falls directly on the waste. Chemical character will be affected by the biological decomposition of biodegradable organic materials, chemical oxidation processes, and dissolving of organic and inorganic materials in the waste leachate's chemical composition will change as the landfill goes through the various phases of decomposition similar to the changes in methane production.

Leachate Treatment and Disposal System

Leachate shall be removed from the drainage and collection system when the leachate level in the landfill interferes with landfill operations or when the unit is subject to assessment monitoring. The operator is responsible for the operation of a leachate management system designed to handle all leachate removed from the collection system. The leachate management system shall consist of any combination of storage, treatment, pre-treatment, and disposal options designed and constructed in compliance with EMA requirements.

The leachate management system shall consist of any combination of multiple treatment and storage structures, to allow the management and disposal of leachate during routine maintenance and repairs.

Standards for on-site treatment and pre-treatment:

   i. All on-site treatment or pretreatment systems shall be considered part of the facility
   ii. The on-site treatment or pretreatment system shall be designed in accordance with the expected characteristics of the leachate. The design may

include modifications to the system necessary to accommodate changing leachate characteristics

iii. The on-site treatment or pretreatment system shall be designed to function for the entire design period

iv. All of the facility's unit operations, tanks, ponds, lagoons, and basins shall be designed and constructed with liners or containment structures to control seepage to groundwater. The ponds, lagoons, and basins shall be inspected prior to use for cracks and settling, and, if leachate is stored in them for more than 60 days, they shall be subject to groundwater monitoring pursuant to this Part

v. All treated effluent discharged to waters of the country shall meet the requirements of national standards for effluent emissions

vi. The treatment system shall be operated by a certified operator

Standards for leachate storage systems:

i. The leachate storage facility must be able to store a minimum of at least five days' worth of accumulated leachate at the maximum generation rate used in designing the leachate drainage system in accordance. The minimum storage capacity may be built up over time and in stages, so long as the capacity for five consecutive days of accumulated leachate, during extreme precipitation conditions, is available at any time during the design period of the facility

ii. All leachate storage tanks shall be equipped with secondary containment systems equivalent to the protection provided by a clay liner having permeability no greater than 10^{-7} cm/s

iii. Leachate storage systems shall be fabricated from material compatible with the leachate expected to be generated and resistant to temperature extremes

Standards for Discharge to an Off-site Treatment Works

Leachate may be discharged to an off-site treatment works that meets the following requirements:

i. All discharges of effluent from the treatment works shall meet the EMA expected standards

ii. The treatment system shall be operated by an operator certified under the requirements

iii. No more than 50% of the average daily influent flow can be attributable to leachate from the solid waste disposal facility. Otherwise, the treatment works shall be considered a part of the solid waste disposal facility

iv. The operator is responsible for securing permission from the off-site treatment works for authority to discharge to the treatment works

v. All discharges to a treatment works shall meet the EMA requirements

vi. Pumps, meters, valves, and monitoring stations that control and monitor the flow of leachate from the unit and which are under the control of the operator shall be considered part of the facility and shall be accessible to the operator at all times

   vii. Leachate shall be allowed to flow into the sewerage system at all times (subject to the town's decision or strategy); however, if access to the treatment works is restricted or anticipated to be restricted for longer than five days, an alternative leachate management system shall be constructed in accordance with world standards

  viii. Where leachate is not directly discharged into a sewage system, the operator shall provide storage capacity sufficient to transfer all leachate to an off-site treatment works. The storage system shall meet the national requirements

Leachate Monitoring

Representative samples of leachate shall be collected from each unit and tested quarterly. The frequency of testing may be changed to once per year for any monitored constituent, if it is not detected in the leachate for four consecutive quarters. However, if such a constituent is detected in the leachate, testing frequency shall return to a quarterly schedule and the constituent added to the groundwater monitoring program. In such a case, the testing frequency shall remain on a quarterly schedule until such time as the monitored constituent has remained undetected for four additional quarters. Leachate and discharges of leachate from units shall be monitored for constituents determined by the characteristics of the waste to be disposed of in the unit. They shall include, at a minimum:

    i. pH
   ii. Annually
  iii. Any other constituents listed in the operator's discharge permit
  iv. All of the indicator constituents
   v. The operator shall also monitor the leachate head within each unit

Time of Operation of the Leachate Management System

The operator shall collect and dispose of leachate for a minimum period of 5 years after closure until treatment is no longer necessary. Treatment is no longer necessary if the leachate constituents do not exceed the wastewater effluent standards (EMA standards).

    If the results of testing of leachate samples in accordance with subsection above show that the leachate exceeds the limits for low risk waste as defined shall:

    i. Notify the Agency in writing of this finding within 10 days following the finding
   ii. Verify the exceedance by taking additional samples within 45 days after the initial observation
  iii. Report the results of the verification sampling to the Agency within 60 days after the initial observation
  iv. Determine the source of the exceedance, which may include, but not be limited to, the waste itself, natural phenomena, sampling or analysis errors, or an off-site source, within 90 days after the initial observation; and notify the Agency in writing of a confirmed exceedance and provide the rationale used in such a determination within ten days after the determination

If, as a result of further testing of the leachate and the background groundwater and analysis, it is determined that the facility leachate exceeds the set limits for low-risk waste, the facility shall no longer be subject to the low-risk waste landfill but hazardous and be subject to the requirements for hazardous waste landfills.

4.50.4   SURFACE WATER SOURCES

According to the hydrogeological report, in the area of the chosen site, there is no significant surface water occurrence within a radius of 0.1 km of the proposed waste disposal facility. Although surface flow does occasionally occur along the non-perennial stream to the southeast of the study area, it is not utilized as a sustainable water source. This stream may, however, act as a transport path for liquid contaminants originating at the proposed facility. The site exhibits a high risk that liquid pollutants from the facility may reach important surface water sources. The site exhibits a slight risk that liquids moving laterally through the soil horizons may eventually emerge at the non-perennial stream to the southeast of the facility.

4.50.5   GROUNDWATER SOURCES

No boreholes were found to occur in the vicinity of the proposed site; however, boreholes will be dug around the landfill to monitor any groundwater contamination. In light of this, it is inferred that the groundwater exhibits a gradient roughly parallel to the regional topography. The site exhibits a high risk that liquids moving laterally through the soil horizons may reach groundwater sources.

4.50.6   EROSION CONTROL DESIGN

There are two types of erosion which need to be avoided: wind and water. In order to provide protection from water the outer slope of the waste disposal site will be provided with storm water channels to avoid possible contamination of storm water with the waste body. Therefore, the water will be navigated around the facility. The surface between the drains should be planted with indigenous vegetation to avoid further erosion via wind. Around the entire perimeter fence a 5 m width should be cleared of vegetation. This is necessary to provide a firebreak. The land operation should strictly practice daily covering of the waste.

4.50.7   METHANE COLLECTION SYSTEM

Various processes take place within the buried waste resulting in generation of various end products. Bacteria in the landfill break down the trash in the absence of oxygen (anaerobic) because the landfill is airtight. A byproduct of this anaerobic breakdown is landfill gas, which contains approximately 50% methane and 50% carbon dioxide with small amounts of nitrogen and oxygen. The composition varies from one disposal site to another and is influenced by waste types. This presents a hazard because the methane can explode and/or burn. So, the landfill gas must be removed via a collection system. To do this, a series of pipes are embedded within the landfill to collect the gas.

The landfill gas represents a usable energy source. The methane can be extracted from the gas and used as fuel. In some developed countries, companies collect the landfill gas, extract the methane, and sell it to chemical companies to power their boilers. The extraction system is a split system, meaning that methane gas can go to the boilers and/or the methane flares that burn the gas. The reason for the split system is that in case of a landfill producing quantities of gases which exceed industrial demand, the gas can be burned into carbon dioxide, water, and other trace gases which are less potent than methane. Alternatively, the gas can be compressed into liquid and sold.

REFERENCES

Bredenhann, L. (2005). *Minimum Requirements for Waste Disposal Landfill, Full5*. The Department of Water Affairs and Forestry, Johannesburg.

Görgens, A. H. M. and Boroto, R. A. (1997). Limpopo River: Flow Balance Anomalies, Surprises and Implications for Integrated Water Resources Management. *Proceedings of the 8th South African National Hydrology Symposium*, Pretoria.

Light, M. P. R. and Broderick T.J. (1998). *The Geology of the Country East of Beitbridge, The Institute of Materials, Minerals and Mining*. Zimbabwe Geological Survey, Harare.

Love, D., Uhlenbrook, S., Nyabeze, W., Owen, R. J. S., Twomlow, S., Savenije, H., Woltering, L. and van der Zaag, P. (2005). Modelling of Hydrological Change for IWRM Planning: Case Study of the Mzingwane River, Limpopo Basin, Zimbabwe. *Abstract Volume, 6th WaterNet/WARFSA/GWP-SA Symposium*, Ezulwini, Swaziland, November 2005, 31.

Masocha, M. (2004). Solid Waste Disposal in Victoria Falls: Spatial Dynamics, Environment Impacts, Health Threats and Socioeconomic Benefits, MPhil, Thesis.

Masocha, M. (2006). Informal Waste Harvesting in Victoria Falls, Zimbabwe: Socioeconomic Benefits. *Habitat International* 30, 838–848.

Minimum Requirements for Waste Disposal by Landfill (DWAF 2nd Edition 1998).

Moyce, W., Mangeya, P., Owen, R. and Love, D. 2006. Alluvial Aquifers in the Mzingwane Catchment: Their Distribution, Properties, Current Usage and Potential Expansion. *Physics and Chemistry of the Earth* 31, 988–994.

Weatherbase: Historical Weather for Beitbridge, Zimbabwe. Weatherbase. 2011. Retrieved on November 24, 2011.

5 Environmental Impact Assessment for Gold Panning

5.1 INTRODUCTION

A company wishes to go into gold panning along the Munyati-Muzvezve riverbed in Kadoma, Zimbabwe. The company secured and registered the mine claim in accordance with the Mines and Mineral Act, CAP 21:05 and seeks to do a small-scale gold panning project. In panning, a miner scoops up a large tray full of sand from a riverbed and agitates it with plenty of water. The heavy gold particles settle to the bottom and the relatively lighter sand can be washed away. Gold panning and associated activities, as specified in Section 97 of the Environmental Management Act CAP 20:27 of 2003, require an environmental impact assessment (EIA) and an environmental management plan (EMP).

The company wants to do gold panning and process the gold, which will be sold to registered buyers. The project life span is more than 5 years. Mining operations have not started. Expected jobs at the mine are more than 100 with locals filling 80% of the posts. The mine has the potential to employ more people from the local community as mining operations expand. Workers will be housed at their homesteads, as all will be from the surrounding area. Water for domestic and industrial use is pumped from the riverbed.

5.2 BACKGROUND INFORMATION

Zimbabwe is endowed with abundant mineral resources, and the processing of these resources is critical to the country's economy. The mining sector employs 5% of economically active adults, contributes 4% of the country's GDP and generates 43.5% of the country's export earnings. The industry is dominated by the mining sectors, namely, multinational mining companies, small to medium scale miners, and artisanal miners. Mining and associated activities result in income multipliers arising from both direct and indirect employment. At the national level, the mining sector has direct positive impact through payment of tax revenue. Indirect positive impact arises from income taxes on employment, personal income, profits of local business, major suppliers, and purchase of goods and services.

Gold panning has also negative impacts, specifically, on the environment. The most significant driver of environmental degradation, gold panning, is expected to cause an increase in sediment, an elevation of sulfates entering water bodies, and an introduction of the toxic metal mercury into the aquatic environment. Gold panning in the confluence will also result in restoration of the pools which were destroyed by siltation. According to rehabilitation plans, sand will be transported to various areas of construction which have been identified. This reduces dependence on virgin soil for construction and reduces financial expenditure for constrictors. An integrated development will result in area benefit.

Apart from limited enforcement of, and compliance with, national law, poor resource use practices, a lack of sense of ownership as well as the need to generate livelihoods among users are responsible for the generation of these impacts (Mensah et al., 2015). It is therefore recommended that illegal forms of small-scale resource exploitation, such as gold panning, be formalized, as will be implemented in this project. Furthermore, a continuous and systematic environmental monitoring system must be set up.

5.3   TERMS OF REFERENCE

In January 2007, Zimbabwe's Environmental Management Agency (EMA) amended the Environmental Management Act CAP 20:27 of 2003 to minimize environmental degradation caused by illegal mining. The Act regulations state that an environmental impact assessment (EIA) should be conducted, approved, and certificte issued prior to implementation of the development projects including mining and associated activities. Accordingly, in order to meet the requirements of Section 97 of the Act, the proponent commissioned a consultant to perform an EMP for gold mining and associated activities at the river area. The EMP outlines environmental management practices that ensure that adequate mitigation or protection measures are incorporated into the project design, implementation, and decommissioning phases in order to allow for sustainable gold mining and processing activities.

A meeting was conducted, at which the scope of the study was described. It was agreed that primary data for the report would be based on ground truthing, site scoping, environmental characterization, and desktop research. Secondary data would be provided by way of disclosure of production quantified performance data, supply analysis records, and mine owner/operator history report.

5.4   GOLD-MINING EIA ESSENTIAL INFORMATION

Guidelines and procedures for granting disposal permits and environmental statutory information governing mining and associated activities were obtained from the Zimbabwe EMA and the Ministry of Mines and Mining Development. Preliminary assessments showed that the project would benefit the proponents as well as the communities and the country at large. Often projects disturb the immediate habitat as what usually happens with development projects; this project has high chances of restoration of the riverbeds. Also, in line with the rural district council development plans which include construction of houses at the shopping center and other identified areas, sand from the riverbed will be used. Virgin land will not be disturbed for construction material. At the same time employment opportunities are created as well as market for sale of community products to mine workers.

5.5   GOLD-PANNING PROJECT DESCRIPTION

The company wants to carry out gold panning and processing on the site and achieve beneficiation by mechanical separation method, i.e. separation of gold from sand by use of a revolving drum system. In the process, a large tray full of sand from the riverbed is agitated with plenty of water. The heavy gold particles settle to the bottom and the relatively lighter sand is washed away (Domfe, 2003). No chemicals will be used, and water will be drawn from the riverbed.

The sand, gold, and mud mixture is transferred to the mobile, diesel-run mechanical separator, which is located either on the riverbed or at the river bank. These activities will be carried out during the dry season, as this seasonal work is only practicable when the rivers are in their low water stages.

A generator will be used as the power source. Grade of mined gold is still to be determined, as it varies along the riverbed. The mine operates three 8-hour shifts per day. There is no crushing or chemicals used in separating the gold – only water is used. The water from the holding tank is recycled back to the mill/plant for use. Tailings or sands from the concentration process are stockpiled at a dump and sent for construction development projects.

5.6   GOLD-PANNING EQUIPMENT REQUIREMENTS

The total gold-panning project worth when it starts to operate will be around USD 198 050. The detailed project costs are presented in Table 5.1.

TABLE 5.1
The Proposed Mining Equipment for Gold Panning

Item Number	Item	Model	Condition	Cost (USD)
1	1 Tractor	Meson Ferguson, 60 HP	Used	8 800
2	1 Front end loader	Caterpillar	Used	60 000
3	Pneumatic compressor	1984 Model	Used	9 100
4	Submersible pump	50 HP	Used	5 750
5	3 Ton truck	Toyota	Used	16 000
6	Water tank pipes and fittings		Used	4 000
7	Office block			25 000
8	Concentrate room			4 000
9	Amalgam room			4 000
10	Workshop			11 000
11	Ablution block and bathroom			14 400
12	Gold panning machine		new	36 000
Total				050 000

5.7   GOLD-PANNING EIA OBJECTIVES

Pursuant to guidelines established by Section 97 of the Act, the objective of this EIA are to:

    i. Identify environmental problems that could arise at the riverbed area as result of gold panning and associated activities

    ii. Suggest ways to manage, mitigate, or, where feasible, prevent potentially negative environmental and social impacts

    iii. Recommend strategies to optimize environmental and social management strategies

    iv. Identify ways in which gold panning and associated activities at the river-bed area can comply with the regulatory requirements of the Act

    v. Make an assessment of the expected socioeconomic and environmental implications of gold panning and associated activities

    vi. Protect the environment by adopting an objective approach

5.8   LEGISLATIVE REQUIREMENTS

The Zimbabwe government, as stipulated in the Act, made it mandatory for an EIA to be undertaken for all major development projects that are likely to have negative impact on the environment. Gold mining and associated activities, as specified in Section 97, require EIAs and EMPs. The Environmental Management Agency is responsible for monitoring projects and advising on environmental concerns. Proponent commits to complying with laws that provide for the protection of Zimbabwe's environment, which includes the following:

5.8.1   ENVIRONMENTAL MANAGEMENT ACT CAP 20:27 OF 2003

Under this Act, the following sections are relevant for the study:

    i. Section 97 lists projects that require EIA/EMP undertakings, and this project falls into the category of projects that require an EIA/EMP

    ii. Under Section 98, the developer is required to submit a prospectus to the Director General containing information regarding the EIA and to state whether the project is a prescribed activity

    iii. Under Section 100, the report will be reviewed by the Director General within 60 days from the date of submission, and, if the report is approved, a certificate valid for 2 years will be issued.

5.8.2   PUBLIC HEALTH ACT CAP 15:09 OF 1971

The Act makes provision for:

    i. Supply of suitable water

    ii. Prevention of pollution of water resources

    iii. Sanitation and control of infectious diseases

5.8.3   Pneumoconiosis Act CAP 15:09

This Act provides for the control and administration of persons in dust occupations. A medical bureau is established in terms of this Act to *inter alia* perform medical assessment and record incidences of pneumoconiosis. Workers are issued certificates of fitness to work in dust conditions.

5.8.4   The Atmospheric Pollution Act of 1971

Under the Act, uncontrolled production and disposal of artificial waste are prohibited. The Act provides for the prevention and control of air pollution by noxious or offensive gases, dust, smoke, and internal combustion fumes.

5.8.5   Mines and Mineral Act CAP 21:05 of 1976

The Act calls for mandatory consultation with the occupier or owner of the land on which the project is located. The Mines and Mineral Amendment Bill 2004, Section 157, proposes that miners, as far as reasonably practical, rehabilitate the environment affected by their operations to its natural predetermined state.

5.8.6   Forest Act CAP 19:05

The minister has wide powers that relate to restrictions on import of specified trees, the suppression of tree diseases, as well as noxious and non-indigenous trees.

5.8.7   Hazardous Substances and Articles Control Act of 1971

This Act regulates the dumping of industrial poisons, pesticides, and other dangerous substances and waste, including radioactive material.

5.8.8   Water Act 20:24 of 1998

Part IV of the Water Act makes provision for the control of water pollution and protection of water resources. In Sections 67–71 of the Act, provision is made insuring that water resources management is consistent with the broader national plan CAP 20:27 SI No 2007. The maximum permissible concentrations of chemical constituents in water discharged or disposed of in a Zone 1 or Zone 2 catchment are listed in Table 5.2.

In addition, the water should not contain any detectable quantities of pesticides, herbicides, or insecticide or any other substance not referred to elsewhere in these standards in concentrations that are poisonous or injurious to human, animal, aquatic, or vegetable life.

5.9   PUBLIC CONSULTATIONS

5.9.1   Methodology

Based on the community characteristics and nature of the project, different methodologies were used for stakeholder consultation. Where possible and within the

TABLE 5.2
Maximum Permissible Concentration of Selected
Chemicals in mg/L in Wastewater

Chemical Constituent	Zone 1*	Zone 2**
Cd	0.01	0.01
Cr	0.05	0.05
Cyanide and other related compounds	0.2	0.2
Hg	0.5	0.5
Ni	0.3	0.3
Zn	0.3	0.3
Fe	0.3	0.3
Total heavy metals	1.0	2.0

* Zone 1: Catchment areas in Zimbabwe's Agro-ecological Region 1
**Zone 2: Catchment areas in Zimbabwe's Agro-ecological Regions 2–5

required statutory frameworks (refer to the Environmental Management Act of 1997), it is also desirable to structure the process in such a way that it would address the needs and interests of the stakeholders. With regard to the EIA for this gold-panning project, the following public consultation techniques were used:

i. Meetings
ii. Public notices
iii. E-mails
iv. Telephone calls

The consultant worked very closely with the ZINWA, Sanyati Catchment Committee, and Rural District Council to develop an appropriate program of stakeholder involvement in the Munyati-Muzvezve confluence area. Consultative meetings were held with the D.A to determine the most suitable method of stakeholder engagement. The D.A and Sanyati Rural District Council suggested having multiple community meetings, each at separate places, in order to increase the participation rate. This course was not taken due to time constraints and other alternative approaches.

Phase 1: Identification of Stakeholders

In the first phase of the engagement process, an initial meeting was held with the various key stakeholders and potential partners. At this meeting, the different milestones for the project were discussed and relevant stakeholders identified. The impact and importance of the support of the community chiefs were discussed, and the project was introduced to the D.A's office and the EMA.

The aim of the project and the EIA was explained, and major issues of concern were obtained. The importance of community involvement was explained and recommendations were made by the D.A.

Phase 2: Stakeholder Meetings and Public Meetings

The second phase of stakeholder engagement was through stakeholder and public meetings. The stakeholder meeting was not held, but the chiefs had to address their communities. During the meetings with I & APs:

    i. The project phase was introduced
   ii. Up-to-date summary and findings were presented
  iii. A discussion was held where all stakeholders were asked to raise their issues, concerns, and questions
   iv. Questions were answered by the client and the consultant
    v. All issues were documented

5.9.2   CONSULTATIONS

Public consultations were undertaken in order to capture concerns of the public and other stakeholders who may directly or indirectly be affected by gold-panning and processing activities at the confluence area. This is cognizant of the fact that mining operations and associated activities frequently involve a high degree of environmental disturbance, which can extend beyond the extent of mined areas. The negative impacts of panning can be experienced in the vicinity of the project and far downstream.

5.9.3   SUMMARY OF THE STAKEHOLDERS PERCEPTIONS

Generally, the project was accepted and appreciated by all stakeholders due to the various benefits it would bring to the community. The main concern was the restoration of the area and avoiding erosion and siltation in the riverbed. ZINWA and the Catchment Committee's main worries were about water resources management and preservation. The proponent should ensure that the water quality is not compromised and a license is legally obtained.

5.9.4   OBJECTIVES OF THE CONSULTATIONS

The objectives of public consultations are to:

    i. Determine the sociocultural and economic context within which gold-mining and processing activities take place
   ii. Ascertain the extent to which gold-panning and processing activities areas are likely to enhance or fracture local people's realization of socioeconomic and environmental values
  iii. Gather stakeholders' views relating to perceived biophysical, environmental, economic, and social impacts arising from mining and associated activities
   iv. Gather information from stakeholders regarding their perceptions of gold panning and best management practices

All the above objectives of the consultation were achieved.

5.10   GOLD-PANNING SPECIALIST STUDIES

This section focuses on the three specialist studies which were performed and the methods used in undertaking the EIA process. The details of the methods are included in specific sections of this report. Each specialist study was approached in a distinct and appropriate way to determine and access relevant information. Below are the approaches which were used by the specialists:

5.10.1   AVIFAUNA, FAUNA, AND VEGETATION

To carry out the avifauna, fauna, and vegetation assessment, the following steps were taken: review of literature, data collection, and information analysis. The compiled report included review of results, impact analysis, and mitigation recommendations.

5.10.2   HERITAGE ASSESSMENT

To carry out the heritage assessment, the following steps were taken: aerial photography, review of literature, and site visits. The compiled report included review of results, impact analysis, and mitigation recommendations.

5.10.3   GEOHYDROLOGY AND GEOTECHNICAL ASSESSMENT

To carry out the geohydrological and geotechnical assessment, the following steps were taken: review of literature, site visits, geological survey, geo hydrogeological survey, geotechnical survey, laboratory testing, water samples tests, soil samples tests, map drawing, and data collection and analysis. The compiled report included review of results, impact analysis, and mitigation recommendations.

5.11   STATE OF THE ENVIRONMENT

Ground truthing, site scoping, characterization, and desktop research were used to determine the state of the environment in the area where the riverbed is located. The states of the environment were categorized into biophysical and socioeconomic.

5.11.1   BIOPHYSICAL ENVIRONMENT

5.11.1.1   Geology

The claim is on the riverbed, and the soil has been previously disturbed by illegal gold panners. Rocks of the area belong to the granite/gneiss geology. Sometimes, alluvial gold occurs in old river banks which have been overgrown by grass and trees in the banks of the river.

5.11.1.2   Soil

Soils are sandy and shallow with depth greater or equal to 2 m.

5.11.1.3 Climate

Climate type is the tropical wet and dry, with two distinct seasons: the wet season from October to April and winter from May to September. Temperature is moderate. Prevailing winds are southeasterly.

5.11.1.4 Vegetation

The surrounding vegetation is largely savanna woodland interspersed with grassed drainage lines. There are no known protected tree species in the area.

5.11.2 SOCIOECONOMIC ENVIRONMENT

Mining is a critical sector in Zimbabwe's socioeconomic development and contributes to fiscal receipts through generation of foreign currency and tax revenues. Mining and associated activities result in income multipliers arising from both direct and indirect employment. At the national level, the mining sector has direct positive impact through payment of tax revenue. Indirect positive impact arises from income taxes on employment, personal income, profits of local business, major suppliers, and purchase of goods and services.

The mine will operate three 8-hour shifts daily. A total of 100 workers, including women, will be employed at the mine. 90% of the workers will be from the local community. There is potential to increase the number of workers and local laborers as operations expand. Workers are housed on site and are provided protective clothing. Potable water will be available from an on-site borehole or from the riverbed.

5.12 POTENTIAL ENVIRONMENTAL IMPACTS

The following impacts are likely to be observed and fall broadly into the following categories: geology, biophysical, and socioeconomic.

5.12.1 BIOPHYSICAL IMPACTS

The biophysical impacts include:

   i. Excavation leading to changes in surface soil
  ii. Excavations exposing soil to heat, leading to soil sterility
 iii. Excavations leading to changes in area topography
  iv. Excavations leading to loss of aesthetic values of the landscape
   v. Excavations leading to land instability as new soil profiles are created
  vi. Gold panning removing soil-binding material, leading to sedimentation loading
 vii. Excavations resulting in soil degradation due to soil structure destruction
viii. Excavation resulting in either increased or reduced infiltration recharge
  ix. Excavation process resulting in air emissions from loading and gaseous emissions from vehicles and motorized equipment
   x. Excavation works resulting in soil and water contamination from fuel spills and grease/oils

xi. Waste discards at excavation sites resulting in poisoning of terrestrial/ aquatic habitat and biota

xii. Exposed sulfur-containing rocks interacting with air and water acid mine drainage.

5.12.2 SOCIOECONOMIC IMPACTS

Gold mining results in positive socioeconomic benefits, which include:

i. Gold-panning projects generate employment
ii. Gold-panning projects sustain local mineral-processing industries
iii. Gold-panning projects contribute to fiscal receipts through tax revenues
iv. Sale of gold generates foreign currency for communities and improves their quality of life

However, negative impacts can also arise, and these include:

i. Improper consultation may cause land ownership conflicts
ii. Competes with agriculture for land and labor
iii. Increased incidence of crime, prostitution, STDs, and HIV/AIDS
iv. New communities, far from formal ones, can lead to alcoholism as workers acquire disposable income
v. Excavations and lack of rehabilitation may lead to entrapment of people/ animals in mine trenches
vi. Noise nuisance from heavy machinery may disturb the community
vii. For workers, respiratory diseases may be caused by working in dusty conditions

The summary of the general potential impacts is shown in Table 5.3.

5.13 ENVIRONMENTAL IMPACT ASSESSMENT METHODOLOGY

Mitigation seeks to minimize or eliminate negative impacts and enhance and maximize positive impacts on people and the environment. Impacts are assessed in terms of their significance to the environment with or without mitigation measures (Kumah, 2006). Significance is a limiting factor in formulating mitigation measures and environmental management plans.

5.13.1 KEY

Environmental aspects: impacts that arise when proposed project activities interact with the surrounding biophysical, social, and economic environment.

Significance: importance of the short-term and long-term changes to the receiving environment. Impacts could be positive or negative. Significance

TABLE 5.3
The General Potential Environmental Impacts from Gold Panning

Mining Phase	Activities	Potential Environmental Impacts
Exploration and surveying	Geochemical, geophysical, and airborne surveys Exploration camp housing Vehicle and machinery parks, fuel points, and service bays Access road construction Waste disposal (garbage) Camp sanitation systems	Vegetation removal, damage, and destruction Habitat disturbance due to noise/vibration Disturbance to wildlife and local residents Soil erosion along trenches and transects Demand on local water resources Discharge or spillage of contaminants Contamination of local groundwater by exposed ores Restricted public access
Mine development start-up; sourcing and stockpiling of raw materials	Mine construction Stripping/storing of soil overburden Surveying and leveling of sites for buildings and plant Installation of mine and surface water treatment plants Construction of mine facilities, offices, and roads Construction of storage facilities Landscaping of site Fauna and flora habitat loss and disturbance Reduction in biodiversity on site Potential loss of heritage sites Decreased aesthetic appeal of site Altered landforms due to construction Altered drainage patterns and runoff flows Increased erosion of site area Increased siltation of surface waters Contamination of surface and groundwater by seepage and effluent discharges Increased demand on local water resources Seepage/discharge of acid rock drainage Ground and surface water contamination from seepage and radionuclides site Construction of staff housing, infrastructure, and recreational facilities	Fauna and flora habitat loss and disturbance Reduction in biodiversity on site Potential loss of heritage sites Decreased aesthetic appeal of site Altered landforms due to construction Altered drainage patterns and runoff flows Increased erosion of site area Increased siltation of surface waters Contamination of surface and groundwater by seepage and effluent discharges Discharge of contaminants via mine dewatering activities Increased demand on local water resources Seepage/discharge of acid rock drainage Ground and surface water contamination from seepage and radionuclides Contamination from fuel spills and leakages Increased demand for electrical power

(Continued)

TABLE 5.3 (CONTINUED)
The General Potential Environmental Impacts from Gold Panning

Mining Phase	Activities	Potential Environmental Impacts
Removal and storage of ores and waste materials	Stripping/storing of soil overburden Waste rock stockpiles Low-grade ore stockpiles High-grade ore stockpiles	Land alienation from waste rock stockpiles and disposal areas Disturbance from vehicle and machinery noise and site illumination Acceleration of acid rock drainage through exposure of ores to air and water Increased erosion and siltation of nearby surface water bodies (rivers and lakes) Contamination of local groundwater
Mechanical separation of gold from ore	Agitating ore to release mineral Transport of ore to separator Extraction and preliminary separation	Ground surface disturbance Disturbance due to noise and vibrations Dust and fumes from mine vehicles and transportation systems Discharge of contaminated water Windborne dust and radionuclides Metal vapor emissions from smelters
Beneficiation	Replenishment of refinery plant processes/solutions Stockpiling of waste and final product	Discharge of contaminants to air, including heavy metals, organics, and SO_2 Spillage of corrosive liquids Requirement for electrical power
Transport of final product to markets	Packaging/loading of final product into Transportation	Disturbance due to noise, vibration, and site illumination Dust and fumes from exposed product stockpiles
Mine closure and post-operational waste management	Decommissioning of roads Dismantling buildings Reseeding/planting of disturbed areas Re-contouring pit walls/waste dumps Water quality treatment Fencing dangerous areas Monitoring of seepage	Subsidence, slumping, and flooding of previously mined areas Acid rock drainage from exposed ores Continuing discharge of contaminants to ground and surface water via seepage Fauna and flora habitat loss and disturbance Windborne dust, including radionuclides Dangerous areas that pose health risks and possible loss of life (e.g. shafts, pits)

is rated on a 3-point scale namely: High (H+/H–), Medium (M+/M–), and Low (L+/L–).

Severity: is described in terms of magnitude, extent, duration, and reversibility.

Magnitude: is the absolute or relative change in size or value on an environmental factor.

Low: When absolute value or relative change is hardly noticeable 0%–10%

Medium: When absolute value or relative change is relatively noticeable 10%–39%

High: When absolute value or relative change is highly noticeable 40%–100%

Extent: The area affected by the impact in requisite unit of measurement

Small: When area affected is smaller than 50%

Large: When area affected is greater than 50%

Duration: The time over which the impact will be felt

Short: 1 year or less

Medium: 1–5 years

Long: more than 5 years

Reversibility: refers to permanency of an impact

Impacts may be reversible in several ways such as:

Reversible by natural means

Reversible by human-aided intervention at reasonable cost

Irreversible when human intervention costs are unreasonably high i.e. greater than all conceivable benefits

The criteria for severity shall be described as:

High: when all four severity characteristics – magnitude, extent, duration, and reversibility – are in the upper extreme categories.

Moderate: when at least any three of magnitude, extent, duration, and reversibility are in the medium categories.

Low: the remainder

5.13.2 Probability of Occurrence

The probability of occurrence refers to the probability of a particular impact occurring. This is subject to the limitations of one's professional judgment in environmental science.

Probability in this case is described as either definite or probable.

Since in most cases it has not been possible to quantitatively calculate probability using mathematical permutations and models, only qualitative descriptions have been adopted. The potential impacts are described in Table 5.4.

The following section discusses in detail each impact identified in the EIA under the respective titles and makes recommendations for its mitigation.

5.14 ENVIRONMENTAL IMPACT ASSESSMENT RESULTS

5.14.1 Physiographical, Topographical, and Relief Features

The Munyati-Muzvezve riverbed lies on confluence 18°22'12.48" S, 22°34'44.55" E. It has a very severe and rigorous topography. Its elevation is 932 m above sea level upwards.

TABLE 5.4
Potential Environmental Impacts of Gold Panning

Environmental Aspect	Potential Impact	Significance	Mitigation	Recommendation
Geology	Scarification	L-	Avoid scarring of the bedrock	Reclamation should be ongoing
	Sterile soils/soil loss (where there are trees and grass)	L-	Plant native tree and grass species on degraded land surfaces	Institute topsoil preservation plan i.e. topsoil on working areas should be stripped and stockpiled for reclamation
	Reduced infiltration recharge	M-	Use vegetative or solid barriers to protect dumping areas for overburden.	
	Loss of visual amenity	L-	Profile waste dumps to acceptable height and slope and transferred to areas of construction identified by SRDC	
Air emissions	Dust nuisance	M-	Use bag filters and other dust suppression equipment	Wet problematic sources of dust routinely
	Gaseous emission	L-	Wet, re-vegetate surfaces	Create a full-time environmental monitoring position at the site
	Noise nuisance	L-	Use effective ventilation systems in buildings and work areas	Storage of hazardous materials should be in secure areas and exercise due diligence
			Provide workers with ear defenders	Use machinery e.g. generators with low noise output
Groundwater	Acid mine drainage from heap and dump tailings	M-	Ensure no crushing activities of the ores	Ensure that used water is returned to the riverbed in clean state
	Heavy metal contamination and associated sediments	M-	Institute sediment and run-off control practice	Implement a fuel spill plan and exercise due diligence
Surface water	Loss of spawn areas due to sediments	L-	Institute erosion control practices to reduce lateral flows	Transport all sand waste to areas of construction activities in Kadoma area
	Sediments of surface water due to erosion	L-	Protect the sand piles from wind blowing	

(Continued)

TABLE 5.4 (CONTINUED)

Potential Environmental Impacts of Gold Panning

Environmental Aspect	Potential Impact	Significance	Mitigation	Recommendation
Vegetation	Biodiversity loss	L-	Practice controlled felling of trees	Monitor tree-harvesting levels
	Loss of soil nutrients and trace elements	L-	Revegetate with native tree and grass species	Monitor to ensure dominance of native species in the mining area
	Airborne pollution (dust and particle)	L-	Wet surfaces	Use spray irrigation
Wildlife/ domestic animals	Death due to drinking contaminated water	L-	Implement hierarchical waste management system i.e. minimization, reuse, recycling, and disposal	Increase volumes for reuse and recycling Use waste inventory to track waste Immediately transport tailings from site to prevent unauthorized access and deter animals
Sociocultural issues	Increased alcoholism/ prostitution	L-	Conduct awareness campaigns on STDs and HIV/AIDS for workers	Support community decision-making structures
Revenue generation	Employment creation/ increase in disposable income	M+	Recruit local people including women for non-specialized jobs	Offer employment age groups
	Forex generation	M+	Keep production records and avoid illegal gold leakages	Conduct a cost/benefit analysis of gold production
Workers' health issues	Dust inhalation	M-	Provide workers with appropriate protective clothing	Institute safe operations and accident-reporting structures and protocols.
	Entrapments in shafts	L-	Support shaft walls with props.	Monitor workers' health routinely
	Respiratory diseases	L-	Monitor furtive emissions and wet surfaces with sealants/water	Run education campaigns for workers
	Communicable diseases	L-	Sanitize areas	

5.14.2 WATER QUALITY

The area around the confluence has low population density with low irrigation intensity. In addition, there are no major sources of organic pollution. The absence of
industries implies that there is no pollution loading from this source as well.

5.14.3 SOIL QUALITY

Soil is sandy loam (outside the riverbed) with presence of free acids and likely
occurrence of exchangeable aluminum. From dispersion ratio, it is assumed that
hydraulic conductivity is very low. Electrical conductivity is normal and porosity
is good for drainage. This soil is good for agriculture and horticulture crops, pH
shows a strongly acidic nature, and the organic carbon content is good. In the riverbed, sand ranges between 0.5–2 m deep and clayey, dark soil with gold deposits
is available.

5.14.4 FAUNAL DIVERSITY

The entire land of the proposed project has practically no forest cover. The animal
habitat is concentrated outside the riverbed, while some animals have adapted to the
riverine environment. Many arthropods such as coleopterans, arachnids, and insects
were observed on the isolated forest patches.

5.14.5 SOCIOECONOMIC STUDIES

5.14.6 WATER SOURCE

The people of the affected villages generally use the water from surrounding boreholes. The population generally fishes from isolated pools in the riverbed.

5.15 ASSESSMENTS OF IMPACTS

Based on the project details and the baseline environmental status, potential impacts
as a result of the preparation and operation on the proposed area exist. Project
impacts have been identified.

5.15.1 IMPACTS ON LAND ENVIRONMENT

The major anticipated impacts during the preparation phase are as follows:

  i. Impacts due to excavation operations
 ii. Impacts due to operation of construction equipment
iii. Impacts due to soil erosion
 iv. Impacts due to sand disposal
  v. Impacts due to construction of roads and houses from the sand supplies

5.15.2 IMPACTS DUE TO EXCAVATION OPERATIONS

During the preparation phase, various types of equipment will be brought to the site. These include mechanical separators, batching plants, and earth movers. The siting of this construction equipment would require a significant amount of space. Similarly, space will be required for storing of various other construction equipment. In addition, land will also be temporarily acquired for the duration of project preparation for storing the excavated material before sending for separation. Efforts must be made for proper siting of these facilities (Mineo Consortium, 2000). During the construction phase, there will be increased vehicular movement for transportation of various construction materials to the project site. A large quantity of dust is likely to be entrained due to the movement of trucks and other heavy vehicles. However, such ground-level emissions do not travel for long distances. In addition, there are no major habitations in the project area. Thus, no significant impacts are anticipated on this account.

5.15.3 IMPACTS DUE TO SOIL EROSION

The runoff from the construction sites will have a natural tendency to flow towards the river. For some distance downstream of major construction sites, such as dams, powerhouses, etc., there is a possibility of increased sediment levels which will impede light penetration, in turn reducing photosynthetic activity which depends directly on sunlight. This change is likely to have an adverse impact on the primary biological productivity of the affected stretch of river and its tributaries. The impact is likely to be greater for the smaller rivers/rivulets where large flow is not available for dilution or are seasonal in nature.

5.15.4 IMPACTS ON WATER QUALITY

The major sources of water pollution during project construction phase are as follows:

   i. Sewage from labor camps/colonies
  ii. Effluent from slime dams

5.15.4.1 Sewage from Labor Camps

About 100 workers are likely to congregate during the project construction phase. The domestic water requirement of the employee population is expected to be of the order of 10 000 L/day. It is assumed that about 80% of the water supplied will be generated as sewage.. The biological oxygen demand (BOD) load contributed by domestic sources will be about 23.7 kg/day. Even if the sewage is discharged without treatment, the minimum flows in the river are much higher than this flow; hence no major adverse impacts are anticipated. However, the sewage generated from labor colonies should be treated before disposal. Normally, during project construction, the labor population will be concentrated at 1 location.

5.15.4.2   Effluent from Mechanical Separators

During the preparation phase, at least one separator near the powerhouse site will be commissioned. The total capacity of the separator is likely to be of the order of 120–150 HP. Water is required to wash the mixture. About 0.1 m^3 of water is required per ton of material separated. The effluent from the mechanical separator would contain high suspended solids. The quantum of effluent generated is of the order of 12–15 m^3/hr. The discharge from the separators does not need to be treated before its disposal on land and/or water. The various aspects covered as a part of impact on water quality during project operation phase are:

  i. Effluent from project colony
 ii. Impacts on reservoir quality
iii. Eutrophication risks

5.15.4.3   Effluent from Project Colony

During the operation phase, the cause and source of water pollution will be much different than that during construction. Since only about 10 people will reside in the area in a well-designed colony with sewage treatment plant and other infrastructural facilities, the problem of water pollution due to disposal of sewage is not anticipated. In the operation phase, about 10 people are likely to be residing in the project area, resulting in about 50 L/day of sewage at 5 L/person. Proper disposal measures for sewage are required to be implemented at the project.

5.15.5   Impacts on Reservoir Water Quality

The flooding of previously forested and agricultural land in the submergence area will increase the availability of nutrients from decomposition of vegetative matter. Phytoplankton productivity can supersaturate the euphotic zone with oxygen before contributing to the accommodation of organic matter in the sediments. Enrichment of impounded water with organic and inorganic nutrients will be the main water quality problem immediately on commencement of the operation. However, this phenomenon is likely to last for a short duration of a few years from the filling up of the reservoir.

5.15.6   Impacts on Terrestrial Flora

A few people (20) including technical staff and workers are likely to congregate in the area during the project construction phase. It can be assumed that the technical staff will be of higher economic status, will live in a more urbanized habitat, and will not use wood as fuel, if adequate alternate sources of fuel are provided. However, workers and other population groups residing in the area may use fuel wood for which firewood/coal depot could be provided. The workers may also cut trees to meet their requirements for cooking. Normally in such situations, a lot of indiscriminate use or wastage of wood is also observed. Hence, to avoid felling of trees by the laborers, alternate fuel supply must be provided.

5.15.7   ACQUISITION OF FOREST LAND

A very limited area outside the riverbed would be required for the mining operations. People will be restricted to areas without vegetation.

5.15.8   DISTURBANCE TO WILDLIFE

Based on the interaction with locals, it was confirmed that within the submergence area, no major wildlife is observed. It would be worthwhile to mention here that most of the submergence area lies within the gorge portion. The river acts as a barrier to movement of wildlife even in the pre-project stage. Thus, the creation of a reservoir due to the proposed project is not expected to cause any adverse impact on wildlife movement.

5.15.9   IMPACTS ON WILDLIFE SANCTUARIES

The resuscitation of the river will attract the wildlife back for water consumption and as habitat.

5.15.10   IMPACT ON THE AQUATIC ECOLOGY

During the construction phase of the preparation of the project, a large quantity of building material like stones, pebbles, gravel, and sand would be needed for construction of various project appurtenances. The cumulative impact of this activity may result in increased turbidity levels. Good dredging practice can, however, minimize turbidity. The second important impact is on the spawning areas of cold-water fisheries. Almost all the cold-water fish breed in the flowing waters. The spawning areas of these fish species are found amongst pebbles, gravel, sand, etc. The eggs are sticky in nature and remain embedded in the gravel and subsequently hatch. Any disturbance of the stream bottom will result in adverse impacts on fish eggs. Thus, if adequate precautions during dredging operations are not undertaken, then significant adverse impacts on aquatic ecology are anticipated.

The pools will change the slow-flowing river to a fast-flowing one. The positive impact of the project will be the formulation of a water body which can be used as commercial fish stocks to meet the protein requirement of the region. Since construction of the pools affects the flow of water in the river, the riverbed below the river site gets invariably affected and many a time a long stretch of riverbed downstream of pools gets affected due to reduction in the quantum of water. However, the minimum flow of water required for the maintenance of aquatic flora and fauna, especially fish, must be maintained downstream of the pools at least up to the tail water discharge point. Proper measures for fish conservation and management may be proposed in the EMP report. The restoration of pools also will not affect the water requirement of the population residing in the downstream areas. This population generally depends upon the local streams and springs for drinking purposes and for other domestic uses. There is also no competitive use of water downstream of river for industrial purposes. Therefore, the impact of pooling in the downstream areas is not anticipated.

5.15.11   IMPACT ON THE NOISE ENVIRONMENT

In a water resource project, the impacts on ambient noise levels are expected only during the project construction phase, due to operation of various construction equipment. Likewise, noise due to quarrying, vehicular movement, and excavation operations will have some adverse impact on the ambient noise levels in the area. Since there are no major habitats in the nearby areas of the project site, it is not likely to have any effect in that regard. No major wildlife is observed in and around the project site. Hence, no significant impacts on wildlife are anticipated as a result of excavation activities.

5.15.12   AIR POLLUTION

In a water resources project, air pollution occurs mainly during the project construction phase. The major sources of air pollution during construction phase are due to fuel combustion from various equipment, emissions from various crushers, and fugitive emissions from various sources. The short-term increase in SO_2, even assuming that all the equipment is operating at a common point is quite low. Hence, no major impact on ambient air quality is anticipated. However, a plan for air quality management is required to be formulated especially for the construction stage of the project in which there will be considerable movement of vehicles and operation of various equipment which may impair the air quality of the project area.

5.15.13   IMPACTS ON THE SOCIOECONOMIC ENVIRONMENT

The preparation phase will last for about 2 weeks. The highest number of the labor force is estimated at about 100. During the construction phase, the basic problem will be the management of many workers migrating to the construction area in search of jobs. Those who would migrate to this area are likely to come from various parts of the country having different cultural, ethnic, and social backgrounds. Such a mixture of population has its own advantages and disadvantages. The advantages include exchange of ideas and cultures between various groups of people which would not have been possible otherwise. Due to longer residence of this population in one place, a new culture, having a distinct socioeconomic similarity would develop which will have its own identity.

The availability of infrastructure is generally a problem during the initial construction phase, though the construction workers can be compensated for certain facilities like health and education. The facilities of desired quality are often not made available in the initial stages. The adequacy of water supply, sewage treatment, and housing should therefore be ensured before and adequate measures be taken at the very start of the project.

5.16   ENVIRONMENTAL MANAGEMENT PLAN

The Environmental Management Plan (EMP) will address all aspects which the project will impact on. It should however be understood that the way excavation will be done will be the starting point of rehabilitation. Such an approach is taken because the riverbed environmental status is dynamic between wet and dry seasons and therefore excavation and rehabilitation should be done concurrently to avoid the seasonal complications and challenges.

Table 5.5 highlights the aspects and the paragraphs after the table deeply explains the management plans of the various aspects of the river. The Environmental Management Plan is shown in Table 5.5.

5.16.1 Environmental Management Plan Details

Environmental protection and sustainable development are the cornerstones of the policies and procedures governing industrial and other developmental activities in Zimbabwe. The Ministry of Environment, Water, and Climate Change has taken several policy initiatives and enacted legislations to prevent indiscriminate exploitation of natural resources. One such initiative is the notification on EIA of developmental projects under the provisions of Environment Management Act (1997) making EIA mandatory for selected categories of developmental projects.

Any mining project that requires land disturbance can provide significant economic and environmental benefits. However, the adverse environmental effects of such a project can also be substantial. In order to make the project fully eco-friendly and ameliorate all possible negative impacts on the economy and ecology of the area, the EMP for the proposed project has been prepared based on the findings of the EIA study of the area. The following management measures are suggested so as to ameliorate the negative impacts as well as to enhance the positive impacts.

5.16.2 Biodiversity Conservation and Management Plan

The riverbed flora and fauna, which are very different from the terrestrial type, depend on the water masses and therefore the loss of water may impact the diversity. Based on this understanding, the management plan for the conservation of the biodiversity of the area would require minimum disturbance of the water pools in the riverbed. This is done by ensuring that no plants are cut down and water is quickly returned to the riverbed after use. The part which has river plants is not subjected to sand collection and is to be left as natural as possible.

5.16.3 Fish Management

High river discharge, fast water currents, and want of suitable spawning ground in the lower reaches of the river are conditions that force fish to swim upstream in search of a suitable eco-system in which to spawn. Creation of pools upstream will offer a new habitat for the migratory path of some fishes and improve the survival and breeding of others. No artificial conditions will be required. Pools which have water are not to be disturbed. Water collection and use should not disturb the fish breeding grounds.

TABLE 5.5

Environmental Management Plan for Gold Panning

Potential Impact	Mitigation	Implementing Agent	Surveillance and Monitoring Agency
Scarification Sterile soils/loss of soil Reduced infiltration recharge Loss of visual amenity	Ensure that excavations do not reach the base rock Plant native tree and grass species on degraded land surfaces Use vegetative or solid barriers to protect dumping areas for overburden Profile waste dumps to acceptable height and slope	Gold-mining company	EMA AGRITEX ZINWA Forest Commission
Dust Nuisance Gaseous emissions	Use bag filters and other dust suppression equipment Use effective ventilation systems in buildings and work areas Institute effective preventative vehicle and equipment maintenance policy	Gold-mining company	NSSA MoHCW EMA
Acid mine drainage from spent ore, heap and dump leach operations, tailings and overburden	Use synthetic liners over clay substrate for dump impoundments Institute sediment and run-off control practices	Gold-mining company	EMA
Heavy metal contamination and associated sediments	Institute erosion control practices to reduce lateral flows Institute sediment and runoff control flows	Gold-mining company	EMA
Loss of spawn areas due to sediments Sediments of surface water due to erosion	Institute erosion control practices to reduce lateral flows	Gold mining-company	EMA AGRITEX
Biodiversity loss Loss of soil nutrients and traces elements Air-borne pollution (dust and particulate matter)	Practice controlled felling of trees and re-vegetate Re-vegetate with native trees and grass species Revegetate disturbed areas and wet surfaces	Gold-mining company	EMA RDC AGRITEX
Sterile soils Air-borne pollution (dust and particulate matter)	Cover with topsoil and use manure in rehabilitated areas Adopt rain watering technique when irrigating re-vegetated areas	Gold-mining company	EMA RDC AGRITEX

(Continued)

TABLE 5.5 (CONTINUED)
Environmental Management Plan for Gold Panning

Potential Impact	Mitigation	Implementing Agent	Surveillance and Monitoring Agency
Wildlife/domestic animal kills due to drinking contaminated water in containing ponds, ditches, and ponds	Implement hierarchal waste management system i.e. minimization, reuse, recycling, and disposal.	Gold-mining company	RDC National Parks and Wildlife NSSA
Respiratory diseases	Provide workers with dust masks Monitor furtive emissions and wet surfaces with sealants/water		MoHCW NSSA
Communicable disease	Sanitize working areas	Gold-mining company	EMA NSSA
Injury/entrapment in pits and shafts	Fence off claim perimeter and tailing dumps Create pools for water to accumulate in them	Gold-mining company	MC MoHCW ZRP
Alcoholism/ prostitution	Conduct awareness campaigns on STDs and HIV/AIDS for workers.	Gold-mining company	ZRP NSSA MoHCW
Employment creation/increases in disposable income	Recruit local people including women for non-specialized jobs Keep production records to deter illegal gold mining	Gold-mining company	NSSA ZRP RDC
Forex gain and GDP growth			
River ecosystem disturbance	Ensure that activities are limited to the silted area which has already undergone degradation. That means the effect is only positive and limited to resuscitation of the river	Gold-mining company	EMA

5.16.4 GEO-ENVIRONMENTAL MANAGEMENT PLAN

A geo-environmental management plan is formulated to protect and/or improve the reservoir zone and to provide stability to the reservoir. The following are the mitigation measures suggested for controlling landslides in the project area.

   i. Rock anchoring, carving out of slopes, shot crating, etc. should be planned
  ii. The impact of a landslide on the project could be managed by arresting the potential landslide zones through suitable engineering treatments like retaining walls and afforestation
 iii. Landslide control with coir-geotextile

5.16.5   Landscaping and Restoration of Construction Areas

Once the construction activities of the project are over, it is also proposed to develop nature parks, children's parks, gardens, and other recreation facilities near the project area. During the construction of the main features, like the powerhouse and residential and project roads, various slopes may be disturbed which shall be stabilized using bioengineering measures like benching, terracing, and planting grasses, herbs, shrubs, and trees.

5.16.6   Public Health Delivery System

During the construction period of the project, about 100 workers who are migrant laborers need to get vaccinated against infectious diseases. The identified possible health threats due to mining construction and other peripheral activities as identified in the EIA study were analyzed and suitable measures are suggested for mitigating the threats. Recommendations for regular health check-ups and a testing program for endemic disease is also suggested. Suggestion for health facilities and infrastructure is made and cost estimation for the same is given.

5.16.7   Solid Waste Management and Sanitation Facilities

In addition, during the construction stage, it is expected that about 100–200 people from nearby villages will visit the project site every day for commercial purposes and constitute the regular floating population. This floating population may also generate solid waste. The quantity of waste generated in Zimbabwean cities is reported to be in the range of 0.2–0.6 kg/capita/day. As the major share of the population is labor force in the area, the waste generation factor of 0.3 kg/capita/day has been taken into consideration. The recommended solid waste management system for the project is presented below:

    i. Segregation of solid waste at source
    ii. Storage and primary collection of waste from project colonies, offices, guesthouses, labor colonies/sheds, minor commercial establishments, market, community center, hospitals, workshops, canteen/mess, school, garden, and parks
    iii. Waste transportation mechanism
    iv. Waste storage depots/enclosures
    v. Waste processing and disposal

Administratively, a Solid Waste Management Committee (SWMC) comprising of the project representatives will look after the management of solid waste. The SWMC will be supported by sanitary workers, the number of which may be decided by the SWMC after assessing the work requirement. As indicated in the site of works plan, slime dams will be managed in coordination with other activities.

Basically, two slime dam sites were identified. Sand waste from the separating machine is deposited in the first slime dam until it reaches capacity and then it's drained into the second one (moved by the bulldozer). That means the first slime dam will again start to fill up. It is unlikely that the slime dam will fill up as sand would be collected for construction purposes. As the process proceeds, closer monitoring is ensured so that the slime dams do not reach a level whereby slides can occur. That's less than 20 m in height. To ensure this, the following steps are taken:

i. Leveling and slope stabilization works are implemented by use of bulldozer
ii. Indigenous trees are planted around the slime dam
iii. Water is sprayed on the slime dam periodically so that the sand is not blown around causing a pollution nuisance
iv. Sand required by the communities is immediately dispatched to the area of need and construction works

The basic arrangement is that the sand from the riverbed would be immediately transported to areas of housing development. Due to such a process approach, it is unlikely that slimes will fill up or mountains of slime develop. This lowers the potential negative impacts from slime dams. The potential negative impacts are associated with sand blowing and liquid from the slime dam.

5.16.8 ENERGY CONSERVATION MANAGEMENT PLAN

It is estimated that during the construction of the project, which would last for about 1 year, around 100 laborers will be working and they will need cooking fuel To discourage illegal tree felling and removal of fuel wood and timber from the adjoining forests, the contractors will provide subsidized kerosene or LPG to their workers. The contractors will also provide community kitchen facilities. And kerosene oil stoves will be provided by the contractors to the laborers as well as the locals in nearby villages. The distribution of pressure cookers is another attractive option for energy saving.

Infrastructure facilities: In addition to the above, the following infrastructure is proposed to be developed in the resettlement colony:

i. Piped water supply
ii. Community toilets
iii. Sewage treatment facilities and sewage system
iv. Community center
v. Vocational activity center
vi. Avenue plantation and block plantation
vii. Internal roads in resettlement colony

5.16.9 DISASTER MANAGEMENT PLAN

The disaster management plan includes evacuation plans and procedures for implementation based on local needs. These are:

    i. Demarcation/prioritization of areas to be evacuated: Working areas will be demarcated and on completion, the areas are rehabilitated before moving to another working area

    ii. Notification procedures and evacuation instructions: These are to be posted in publicly accessible places for workers and the community

    iii. Safe routes, transport, and traffic control are established

    iv. Shelter areas are designated

    v. Functions and responsibilities of members of evacuation team are documented

The copies of the Emergency Action Plan should also include the inundation map, which would be displayed at prominent locations and in the rooms and locations of the personnel named in the notification chart. Inundation maps will be displayed near the project area and also in the villages in flood-prone zones. For speedy and unhindered communication, a wireless system will be a preferable mode of communication. Telephones would be kept as backup, whenever required.

5.16.10 MAINTENANCE OF AIR, WATER, AND NOISE

The EIA reports that the quality of water of the river is reasonably good. At present, no developmental activities such as industries etc. are going on upstream of the catchments; therefore, any probabilities of water quality degradation are minimal. Also, there are few human habitations draining refuse into the river which could charge the nutrient status of the river waters, thereby degrading the aquatic ecosystem. However, the project authority should take effective and proactive measures to ensure that such activities would not be carried out in the upstream catchment, which may bring about water quality degradation in the future as well. Necessary financial outlay for establishing water quality testing has been documented in the Environmental Monitoring Plan.

In case of mining projects, air and noise pollution mostly occur during the construction period when different project-related activities like excavation, use of diesel generators, muck disposal, etc. are undertaken. It is therefore recommended that necessary preventive measures be taken during all those activities that can lead to air and noise pollution. The various crushers need to be provided with wet scrubbers to control dust generated while crushing the stone aggregates. It should be made mandatory for the contractor involved to install cyclone separators/scrubbers in crushing plants. During the execution of the project, due care has to be exercised to minimize the exposure of workers to excessive noise. As far as possible even consideration is to be given to locate the site office, stores, etc. in the minimal noise locations. Appropriate safety measures for workers (e.g. protective equipment for workers like ear protectors, ear muffs, ear plugs/defenders) to protect from high noise levels need to be adopted.

5.16.11 ENVIRONMENTAL MONITORING PLAN

Monitoring becomes essential to ensure that the mitigation measures planned for environmental protection function effectively during the entire period of project operation. It will also allow for validation of the assumptions and assessments made in the present study. An environmental monitoring cell (EMC) will be formed in order to assess and review the progress of the various mitigation measures suggested in the EMP. The committee will sit at predetermined intervals for verifying progress and reporting the same. The project authority shall depute a senior officer to coordinate with the monitoring committee.

The project authority will engage a neutral agency or organization for supervision and monitoring of the environmental management components as discussed below. The project authority will also depute a full-time senior officer to observe and coordinate the progress of the environmental management activities. The independent supervising agency will work closely with the project environmental cell, carry out the necessary laboratory analysis and collection of data regarding the progress, and prepare a progress report every two months to be presented to the Monitoring Committee. For any major comments or obstacles, the independent agency may call a meeting where representatives from independent agency, project authority and environmental committee will be present and any issue may be discussed in the meeting. The progress reports will cover:

   i. Progress of panning area, treatment works, fish management
  ii. Status of protection measures (e.g. sausage/gabion walls, etc.) at the dumping and quarry sites
 iii. Leveling and slope stabilization works at dumping sites
  iv. Status of afforestation/turfing works on the dumping/quarry sites.

Based on the findings of the EIA study in various

EMPs, the important parameters are: biodiversity conservation and management, public health delivery system, fish management, restoration of dumping sites, landscaping and restoration of construction area, and green belt development.

The following will be monitored:

Surface water quality: should be monitored twice a year. About 6 samples need to be analyzed. This analysis shall be performed up to the commissioning of the project.

Air quality: will be monitored by the cell quarterly, in terms of sulphides abd oxides (SOx, and NOx).

Additionally, the status of afforestation programs, changes in migration patterns of the aquatic and terrestrial fauna, soil erosion rates, slope stability of embankment, etc. should be studied every 5 years until the commissioning of the project.

Identification of water-related diseases, sites, adequacy of local vector control and curative measures, status of public health are some of the parameters which should be closely monitored once every two years with the help of data maintained in the government dispensaries/hospitals. In addition to the above following parameters will also be monitored by the EMC:

   i. Status of protection measures (sausage/gabion walls etc.) at the dumping and quarry sites

ii. Leveling and slope stabilization works at dumping sites
iii. Status of afforestationon the dumping/quarry site

5.17   GOLD MINE DECOMMISSIONING

Sustainable development, with its premise of equity amongst the present and future generations, requires institutions of planning and execution of post-closure activities that address environmental impacts arising from gold milling and processing activities in order to bequeath to posterity a landscape that will support life-giving ecosystems and optimize activity. Mine closures are governed by:

5.17.1   NATIONAL ENVIRONMENTAL POLICY

The government requires mining companies to develop mine closure plans aimed at rehabilitating the site and surrounding areas affected by mining activities to the extent possible, and set aside resources for their implementation, so as to reduce long-term negative environmental effects.

5.17.2   MINE AND MINERAL ACT (CHAPTER 21:05) REVISED VERSION 1996

This act calls for mandatory cooperation with the occupier or owner of the land in which the mine is located. A quittance certificate is only issued with the consent of the landowner.

5.17.3   MINES AND MINERALS AMENDMENT BILL, 2004

5.17.3.1   Section 157

proposes that miners, as far as reasonably practical, rehabilitate the environment affected by the mining operation to its natural or predetermined state or a land use that is consistent with the principle of sustainable development.

5.17.3.2   Section 159

proposes that mines should assess their environmental liability and establish a fund that shall be used for the purpose for rehabilitation or management of negative environmental impacts.

Decommissioning activities should cover both physical and social aspects as shown in the following section:

5.17.4   PHYSICAL ASPECTS

The physical aspects include:

i. Ensure that deserted mine activities are in a safe non-polluting state
ii. Institute natural restoration of site through refilling and replanting of viable vegetation

iii. Ensure detoxification of all gold plant tailings prior to use in mined out slopes as backfill
iv. Profile waste dumps to acceptable height and slope
v. Institute sediment and runoff controls
vi. Rehabilitate areas disturbed during construction and operation
vii. Remove all technical elements with a limited lifetime
viii. Areas formerly covered with concrete or any other sealants must be excavated, regarded, and recovered with topsoil so that stable native soil remains

5.17.5 SOCIOECONOMIC ASPECTS

The socioeconomic aspects must address:

i. Redundancy packages must be worked out for employees based on the experience and post held
ii. Prepare workers for decommissioning to reduce job dependency and income vulnerability (Provide skills evaluation, exercise, and, possibly, ensure certification)

The summary of each environmental impact is given in Table 5.6. Each environmental issue identified under each assessment class aspect is characterized in Table 5.6.

5.18 DETAILS OF TAILINGS MANAGEMENT

5.18.1 POND STORAGE

Tailing ponds are areas where the water-borne refuse material (mining tailings) is pumped into a pond to allow sedimentation of solid particles from the water. The pond is generally impounded with a dam, and known as tailings impoundments or tailings dams. The ponded water is of some benefit as it minimizes fine tailings from being transported by wind into populated areas where the toxic chemicals could be potentially hazardous to human health; however, it is also harmful to the environment. Tailing ponds pose a danger to wildlife, such as waterfowl, as they appear to be a natural pond but can be highly toxic and harmful to the health of these animals. Tailings ponds will be used to store the waste made from separating minerals from rocks.

Large earthen dams will be constructed and then filled with the tailings. Exhausted open pit trenches which are deep will be refilled with tailings. In all instances, due consideration must be made to contamination of the underlying water table, amongst other issues. Dewatering is an important part of pond storage, as the tailings are added to the storage facility the water is removed – by draining into decant tower structures. The water removed will thus be reused in the processing cycle. Once a storage facility is filled and completed, the surface can be covered with topsoil and revegetation commenced where the water does not flow. However, unless a non-permeable capping method is used, water that infiltrates into the storage facility will have to be continually pumped out into the future slime dam. The biggest

TABLE 5.6

Assessment of Each Potentially Significant Impact during Mining Operation

Environmental Aspect	Nature of Impact	Extent of Impact	Duration of Impact	Probability of Occurrence	Degree of Reversal	Degree of Irreplaceable Loss	Mitigation Measures	Level of Significance after Mitigation
Dust pollution	Negative	Local	Long-term	High	Medium	Low	Use of water for suppression	Low
Groundwater pollution	Negative	Sub-regional	Short-term	Low	High	Medium	Dilution of water	Low
Waste scattering	Negative	Local	Long-term	High	Medium	Medium	Transfer sand to construction site	Low
Soil contamination	Negative	Local	Short-term	Low	Medium	Medium	Removal of sand waste from site	Low
Storm water	Negative	Local	Short-term	Low	Medium	Low	Collect in pools	Low
Runoff from the site	Negative	Local	Short-term	Low	Medium	Low	Use of collection system	Low

danger of tailings ponds is dam failure. Tailings ponds can also be a source of acid drainage, leading to the need for permanent monitoring and treatment of water passing through the tailings dam.

5.18.2 Dry Stacking

Tailings do not have to be stored in ponds or sent as slurries into rivers. An alternative is dewatering tailings, using vacuum or pressure filters, so they can then be stacked. This saves water, reduces impacts on the environment in terms of space used, leaves the tailings in a dense and stable arrangement, and eliminates the long-term liability that ponds leave after mining is finished.

5.19 CONCLUSION AND RECOMMENDATIONS

5.19.1 Conclusion

The gold sector is critical in terms of revenue generation for countries that rely on mining. Environmental management and monitoring are essential in order to promote safe mining, handling of chemicals, and post-closure of the mine. This is crucial in the case of riverbed panning due to the potential negative effects on water streams.

5.19.2 Recommendations

Going into the future, alternative technologies for gold processing in riverbeds must be implemented in line with the Minamata Convention. Continuous training of the artisanal and small-scale miners in safe mining methods is recommended.

REFERENCES

Domfe, K. A. (2003). Compliance and Enforcement in Environmental Management: A Case of Mining in Ghana. *Environmental Practice* 5(2), 154–165.
Kumah, A. (2006). Sustainability and Gold Mining in the Developing World. *Journal of Cleaner Production* 14(3), 315–323.
Mensah, Albert K., Mahiri, Ishmail O., Owusu, Obed, Okoree D., Mireku, Ishmael Wireko and Kissi, Evans A. (2015). Environmental Impacts of Mining: A Study of Mining Communities in Ghana. *Applied Ecology and Environmental Sciences* 3(3), 81–94.
MINEO Consortium (2000). Review of Potential Environmental and Social Impact of Mining.

6 Environmental Impact Assessment of a Scrap Metal Smelting Plant

6.1 INTRODUCTION

Huge amounts of scrap metals are being generated on a daily basis and can pose environmental damage if not recycled (Danielson, 1973). A company in Willowvale, Harare, Zimbabwe intends to embark on a smelting plant project of solid metallic components in the area. The company has identified the need to smelt scrap metal for molding purposes. The smelting plant will melt the scrap metal for use by other individual companies. The smelting of metals is critical in the sense that most of the waste metal lying idle will be reused, thereby promoting sustainability.

Since the Environmental Management Agency (EMA) requires that all development projects should have an Environmental Impact assessment (EIA) report before implementation, this report fulfils that requirement. This EIA document describes the environmental impacts identified, outlines mitigatory measures, and points out how the proponents can manage the negative environmental impacts caused by the operations of the project. The environment in the project area as divided into the biophysical environment, the economic environment, and the sociocultural environment needs careful analysis so that minimal negative impacts are realized or, rather, avoided at all costs. The EIA report is an important tool in the project planning, development, and decision-making processes. The report should provide accurate, useful, and relevant information so that the environment is protected and that the proponents of the project conform to the dictates of the law for sustainable development.

6.2 NECESSITY AND SCOPE OF THE ASSESSMENT

A scrap metal smelting plant has some negative environmental impacts associated with its development. There are air, land, water, and noise pollution issues, as well as chemical waste issues. It is therefore the main objective of this report that these issues are addressed to minimize environmental damage in accordance with the laws regulating the establishment of such developmental projects.

In order to have an appreciation of the project area, the consulting team must make visit to relate the current economic activities taking place in the area of the proposed project. Fuerthermore the activities of the new project on how they would impact the immediate environment and in some ways affect distant environment through fluvial processes. The data collection process included the following aspects:

    i. Emissions
    ii. Traffic or vehicular movement patterns
    iii. Land use patterns
    iv. Noise pollution
    v. Waste management (solid and effluent)

6.3   IMPACT IDENTIFICATION

After an analysis of the collected data, the consultancy conducted an impact identification process of the project. This entailed identifying the potential negative environmental impacts of the project from project design up to the decommissioning stage. A Six Activity Model was used for the production of the EIA report. This model has been used by the consultancy over time and consists of the following steps:

    i. Current environmental status description
    ii. Acquisition of relevant standards or guidelines
    iii. Determination of potential impacts of the proposed project
    iv. Impact prediction
    v. Determination of impact significance
    vi. Determination and incorporation of mitigation measures

6.4   ENVIRONMENTAL MANAGEMENT PLAN

An environmental management plan (EMP) is a tool that is used for checks and balances in the life of the smelter plant. In the EMP, the potential environmental impacts and mitigatory measures are identified and evaluated to indicate the levels of severity of the impacts to the environment. The EMP also provides the mitigatory measures that address the negative impacts.

6.5   SMELTING PLANT EIA PROCESS

The study progresses in the following manner: provides a description of the site, describes the proposed components of the project, reviews the policy context in which the environmental impact assessment has been carried out, presents an environmental management plan for the project, and gives a summary and conclusion of the EMP.

6.6   SMELTING PLANT BASELINE STUDY

The smelter plant sits on approximately 2 hectares of land. The heavy industrial area is to the North of the plant.

6.6.1   AREA HYDROLOGY

The area on which the smelter is to be established is generally flat and covered with a variety of grass and shrub species. The soils are generally black clays, and the drainage is poor in these soils. The area is vlei-like, but this is seasonal, as waterlogging only occurs during the rainy season. Reeds in the stream channel suggest that there water present in the area, which can be described as a wetland, though it is not a fully-fledged wetland.

6.6.2 Vegetation

The project site surroundings are covered by riverine grass; no woody species were identified. Grass species compatible with waterlogged areas are prevalent on the project site, and the entire setup looks swampy. These grasses are spread within the project site surroundings and are thriving because of the rainy season. The common grass species present are the *Sporobolas spiramidalis*, *Paspalam puverli*, *Typha capensis*, and the *speras*.

There are also a couple of reeds on the premises, indicative of intermittent high water and a swampy area. It is important to note that during the summer season, the project site displays an abundance of water but will decrease in winter due to less rainfall. These may also be attributed to some pits which had been left open and then later disused. The soil type is black and deeply saturated with water, the soil grains adhering to each other. These clayey soils leave small air spaces within them leading to soil with generally poor aeration.

6.6.3 Climatic Conditions

The savannah climate prevails in the project area, as Harare lies in Region 2 of the agro-ecological regions of Zimbabwe. Rain falls from mid-October to April. The peak rainfall period is between December and February. Harare receives about 860 mm of rainfall per annum. Temperatures are highest in October and November, whilst the lowest temperatures are experienced in June and July. The average temperature is 22°C, with a diurnal range of 14°C in summers while in winter it is 22°C. Evaporation rate is high in summer months because of the high rainfall and high temperatures; it is conversely low in the winter months due to low rainfall and temperatures.

Air quality monitoring still presents a challenge in the country due to the lack of modern and sophisticated testing equipment. EMA is said to have recently acquired some testing equipment, but there is limited use of the equipment. Potential air quality violations are based on the use of known combustible fossil fuels; particulate dust emissions (soot and fine ash) are also introduced to the atmosphere together with the smoke. The air in the industrial area is characterized by the presence of the identified gases and the particulate matter arising from the combustion of the fossil fuels used in industry. Between these pollutants and the general wind movement from east to west, the air quality in the western parts of the industrial area is poor.

6.6.4 Water Quality

Harare is part of the Lake Chivero catchment area and several rivulets and or ephemeral streams pass through the city's industrial area into the Hunyani River system.

The water quality and testing laboratory managed by the EMA conducts water quality testing. District environmental officers and provincial officers submit water samples for periodic testing at the lab. In most cases, water quality testing is done when an environmental accident happens which introduces pollutants to the river channel systems. Harare City Council (HCC) has the sole mandate to procure and provide potable water for use in the area under the jurisdiction of the city council. The water is treated at Morton Jaffray Water Treatment Works near Norton. The water is then delivered to Harare through a network of pipelines and there are pressure booster stations at strategic places in and around Harare. The water is delivered to homes and industry by the HCC.

6.7   DESCRIPTION OF THE SMELTING PROJECT

After the government banned the export of scrap metal, the company currently doing business in the industrial area recognized a business opportunity and decided to establish a steel smelting plant. The government observed that there was so much scrap metal being exported to the Republic of South Africa, and that smelting of scrap metal in-country has both economic and environmental benefits: job creation, energy savings, and less reliance on metal imports. On the environmental side unwanted scrap metal pollutes the land. By smelting it, the environment is rid of unused and unwanted metal which is then recycled and stored in the form of ingots for use in the future. And because rusty metals are a source of tetanus, (a disease that causes muscle spasms and rigidity), recycling scrap metal promotes hazard-free playing environments for children and safer workplaces, where injuries from unwanted scrap metal are reduced. The metallurgical process in industry is shortened as part of the metal is readily available in the form of scrap metal.

6.7.1   PLANT LOCATION

The smelting plant shall be located in the Willowvale industrial area. This area is home to existing heavy and light industrial works from metal works to plastics engineering. The smelter plant brings an advantage to metal engineering firms in that the smelter products may be readily acquired by the industries in the locality of the smelter plant.

6.7.2   SMELTING PROCESS

Smelting of the metals or ore requires a furnace designed by engineers according to the requirements of the project proponent. The smelting plant shall operate on a 24-hour basis and shall have routine planned maintenance programs to ensure safety and efficiency. The processing of scrap metal requires the following components to be fully operational:

6.7.3   TRANSPORTATION OF SCRAP METAL

A ready stockpile of scrap metal is a general requirement in order to keep the furnace operating. When the furnace is turned on, certain temperatures need to be maintained to avoid the collapse of the furnace structure. Starting the furnace and getting it to the required temperature is a slow and expensive process. Trucks shall be used to ferry the scrap metal to the plant. The daily expected delivered tonnage shall range between 120 and 150 tons of scrap metal. A weigh bridge shall be put in place to weigh the trucks before they proceed to the stackyard.

6.7.4 STACKYARD FOR SCRAP METAL

The stackyard is the place where all the scrap metal is deposited. The area shall have a concrete top of an excavated and gravel compacted base to accommodate the heavy trucks. The scrap metal shall be sorted according to types and stacked in rows with clearance in between them to allow vehicular and wheelbarrow movement. A scrap metal hydraulic compactor shall be introduced to enhance the storage of the scrap metal.

6.7.5 RECEIVING SECTION

Wheel barrows and or hydraulic carts shall be used to move the scrap metal from the stackyard to the receiving (intake) section of the smelter where it is then fed into the furnace. The receiving section has a ramp about 1 m in height elevating from the ground level to the top of the smelter. The scrap metal is then fed into the smelter from the top.

6.7.6 SECTION FOR SMELTING

The smelter shall have 2–3 m of its total height encased in concrete. There will be two chambers on the furnace and these will receive approximately 1000 kg (1 ton) of scrap metal each. The chambers will be tightly closed, and the scrap metal is subjected to highly pressurized heat for one and quarter hours (1 hour and 15 minutes). The scrap metal starts to melt into molten liquid where 90% of the scrap metal is recovered while 10% is lost as waste or burnt completely. The burnt component would be the rust and other materials like plastic washers and paint that come together with the scrap metal. Hydraulic mechanisms pour the molten liquid into melds that are 6 cm wide and about 1.0–1.5 m in length. When cooled the bars are then referred to as ingots which weigh about 23 kg and are stacked in a shed for further processing into various forms of steel. Such a process is envisaged to bring high yields. The plant will use 2000 MV of power and a dedicated power line from a nearby Zimbabwe Electricity Transmission and Distribution Company (ZETDC) will supply the plant with power. The dedicated power line is an essential component of the project because electricity interruptions can cause damage to the plant. The plant also uses coking coal for the smelting process.

6.7.7 LABOR FOR SMELTING PLANT

The smelting plant requires a 45-man strong labor force of both skilled and unskilled personnel for the plant to be fully operational. The labor force may be categorized as:

    i. Management and administration 6
    ii. Skilled and semi-skilled 21
    iii. General and unskilled 18

The staff compliment above has various skills and functions in the whole process chain.

6.7.8 BY-PRODUCTS OF THE SMELTING PLANT

Modern smelters are efficient both in energy and process efficiency. Smelting of scrap metal, unlike the smelting of mineral ore, produces little or no solid wastes. The impurity level in the scrap metal is very low and takes the form of paints, rust, and other components such as plastic and paper washers. These completely burn out or produce little waste. Less than 10% of the total weight charged into the smelter is lost in the process.

6.7.9 EXISTING INFRASTRUCTURE

The smelting plant has various existing infrastructure.

6.7.9.1 Water Supply

The HCC has a water supply infrastructure to supply the smelting plant. The water supply is the City Council's responsibility as it has the lawful mandate to be the sole operator and supplier of potable domestic and industrial water through its water supply infrastructure. The water is for cooling and cleaning purposes. The water from the cooling system is recoverable on site. A storage tank shall be constructed at the site to ensure availability of water at the plant in the event of failure in the city supply line.

6.7.9.2 Sewage System

Liquid waste water from the ablution block, canteen, and spillage from the smelter plant shall be disposed of through the City Council sewage system. The volume of liquid waste matter is low due to the small labor force at the plant. The smelter discharges minimal water through accidental spillage where this water collects on the floor and gets dirty.

6.7.9.3 Electricity

The ZETDC has electricity grids to supply the various industries in the area. There is a ZETDC substation within a short distance from the proposed plant. The company is presently serviced by an 11 KV line, but this line shall be upgraded to supply 2000 MV to the smelter. The company shall also install a standby generator should there be interruption in electricity supply.

6.7.9.4 Road Access

The area where the plant is located is serviced by two major roads namely Willowvale and High Glen Roads. It is unlikely that traffic to and from the plant will cause any congestion, as only 3–4 trucks are expected at the plant on a daily basis to offload scrap metal.

6.7.9.5 Administration Block

There is an existing building on the site which currently houses administrative offices. This building was left by the previous occupant. The company is in the process of upgrading the building. Another building shall be erected to house the canteen, workers' changing room, and toilets.

6.8 LEGISLATION AND REGULATORY FRAMEWORK

This section identifies policies and legislation that are likely to affect or have implications for the smelting plant. The discussion only focuses on those sections in the different acts which relate to the smelting plant and its operations, the local authority, and related environmental issues. With the enactment of the Environmental Management Act, environmental impact assessments are now a requirement for new projects and project modifications, such as the smelting plant. This EIA will review the necessary acts and make sure that the activities at the smelting plant have complied with the standing regulations and legislation.

6.8.1 ZIMBABWE EIA POLICY AND GUIDELINES

Zimbabwe has endorsed the concept of sustainable living and use of its natural resources and intends to use the EIA Policy to support those concepts. The then-Ministry of Mines, Environment and Tourism published the EIA Policy in 1997. It is also accompanied by a set of general and sectored guidelines for carrying out EIAs. Volume 8 of these EIA guidelines deals with mining projects, under which this development would fall.

The Environmental Management Agency (EMA) administers the EIA Policy. The Minister is the lead authority and the Director of the EMA has been delegated the responsibility of overseeing the processing and permitting of EIA submissions. There are three sequential stages in the process:

6.8.1.1 Proposal Development

This involves the preparation of a prospectus early in the planning of a project, informing EMA that an activity is being considered which may warrant the preparation of an EIA. EMA, using a list of prescribed activities and screening guidelines, then assesses whether an EIA is required or not.

6.8.1.2 EIA Preparation and Review

If an EIA is required, the project proponent will then be required to prepare the report. The policy and guidelines give the requirements and procedures for carrying out the EIA studies. This includes scoping, public consultation, preparation of Terms of Reference, specialist EIA studies and contents of the EIA report. EMA reviews the EIA report and may seek additional information. When the review is complete, the EMA Director may decide to either authorize or refuse "EIA

Acceptance". This will not be considered complete until all appeals, if any, have been dealt with.

6.8.1.3   Implementation

Once an activity has been granted EIA acceptance, the proponent may then proceed to the permitting authority to undertake the activity. The EIA process during implementation and operation will include monitoring and auditing. The EIA process is now mandatory for all projects listed in the First schedule of the Environmental Management Act (Chapter 20: 27) among which the smelting plant, being in the form of an urban infrastructure operation, is inclusive. Since being passed, the Environmental Management Act has enforced the EIA Policy, and all the prescribed activities need a full EIA to be carried out before operation commences.

6.8.2   Environmental Management Act (Chapter 20:27)

The Environmental Management Act that repealed the Natural Resources Act was passed by Parliament and became effective on 17 March 2003. The Act harmonizes environmental management by combining the Natural Resources Act, the Atmospheric Pollution Prevention Act, the Hazardous Substances and Articles Act, and the Noxious Weeds Act. The following parts of the Act are relevant to this study:

> *Part XI:* This part of the Act requires developers to take all reasonable measures to prevent or, if prevention is not possible, to mitigate any undesirable effect on the environment that may arise during the implementation of a project EIA report. The Act details that the developer shall report to the Director General any measures to address environmental issues, unless the measures have been described in an EIA report.
>
> *Part IX:* This part of the Act deals with environmental quality standards and pollution controls including compliance, discharge of pollutants, penalties for pollution, pollution prevention, pollution permits, waste management, and licenses for waste. It also lists the duties of local authorities in relation to pollution.

6.8.3   Effluent and Solid Waste Disposal Regulations

Statutory Instrument 6 of 2007 Section 12 places the responsibility of every generator of wastes to prepare, implement, and adhere to a waste management plan which shall consist of an inventory of the waste management situation and specific goals for the management of the wastes.

6.8.4   Environmental Impact Assessment and
Ecosystems Protection Regulations

Statutory Instrument 7 of 2007 Section 13 spells out how the EMA will carry out bi-annual environmental audits to ensure that all projects are in compliance with ecosystem protection regulations; it also places the responsibility on developers to submit quarterly environmental monitoring reports to report on the environmental state of their projects and activities.

6.8.5 REGIONAL, TOWN, AND COUNTRY PLANNING ACT (CHAPTER 29:12)

Part IV of the Act gives local authorities the power to prepare master and local plans for their areas. One of the conditions in the Act (Section 14.2) is that the master plan should include measures "for the conservation and improvement of the environment… and indicate parts of the planning area which are of high scenic value and should be protected." (Environmental Management Act, 2007) In this regard, the smelting plant is covered by two subject plans: the Harare Combination Master Plan is the general plan covering the whole city, and Local Development Plan 36 is more specific on the project site and surrounding areas.

6.8.6 HARARE COMBINATION MASTER PLAN

The Harare Combination Master Plan was prepared in terms of the provisions set out in Section 14 of the Regional, Town, and Country Planning Act (1996). As such, the Harare Combination Master Plan sets out provisions and guidelines for the preparation of Local Development Plans.

The following issues are relevant and must form the basis to the preparation of the Local Development Plan 36.

i. **Population (Growth and Migration)**
   To regulate the population distribution within the plan area with anticipated average economic opportunities and growth during the plan period.

ii. **Infrastructure and Land Development Potential**
   Ensure proper maintenance and optimum use of land and existing public infrastructure.

   Where feasible permit smaller subdivisions in existing low-density housing areas that are on reticulated sewage disposal and water supply systems as well as those on septic tanks (subject to satisfactory porosity tests and performance). Connect all areas which will gravity feed to the existing sewage disposal system and, where feasible, connect parts of existing residential areas to the existing sewage systems by means of pump stations and other technical measures to permit sub-divisions to encourage more intensive use of residential land. Where feasible, extend and upgrade sewage reticulation capacity in certain built-up areas in order to encourage intensified use of land.

   Strike a balance between the need to provide some land for urban expansion and associated infrastructure and the pressing need to safeguard prime agricultural land, areas of natural beauty, and recreation areas. Provide and upgrade the necessary public infrastructure on a cost recovery basis in order to accommodate new developments and encourage the intensive use of land throughout the planning area. Maximize use of land in existing low-density residential areas by permitting a harmonious mix of technically feasible smaller subdivisions to enable the development of all types of residential layouts and housing types without devaluing the environment.

iii. **Traffic and Transportation**

Ensure that the traffic and transportation system serves land use effectively, ensuring that the spatial structure of the planning area remains unobtrusive and that persons and goods can move or be moved from place to place rapidly, safely, conveniently, and economically within and throughout the planning area. Develop an integrated traffic and transportation system that serves all areas efficiently. Provide transportation routes which form part of an efficient circulation and distribution system, each link fulfilling its purpose as part of a planned hierarchy.

Provide for different transportation modes which complement each other in terms of routes, vehicles, and terminus. Provide and promote the use of integrated bicycle routes in urban areas, including the necessary grade separations at road intersections and rail crossings. Where possible, such routes should provide direct, safe, and independent linkages between residential and employment centers. Reduce traffic accidents by innovative use of modern traffic restraint and management techniques and principles.

Endeavour to develop a hierarchical transportation system by upgrading and managing some existing roads and developing new linkages lading to future integrated metropolitan structures in order to serve the planning area efficiently. Aim to achieve an efficient traffic flow and resolve likely conflicts between vehicles and vehicular/pedestrian traffic movement by adopting and incorporating up-to-date traffic management techniques.

iv. **Industry and Employment**

Where appropriate, encourage decentralization of industrial activities to the smaller urban centers. Identify areas/sites in the established urban centers where a regime of industrial activities can be permitted e.g. designated enterprise zones and/or planning zones with minimum development restrictions. Introduce a range of planning and other incentives likely to attract industries to establish in such centers/localities.

Create employment opportunities in the peripheral centers by designating selected urban areas "growth points" and by encouraging decentralization to such centers through a range of incentives. Create conducive conditions and provide accessible sites for use by small-scale emerging business persons, and designate areas where income-generating activities can be combined with residential use on individual stands in order to mop up the existing abundant marginal labor resource and as a means to encourage and tap self-initiative of individuals within the population.

v. **Environment**

Ensure conservation and protection of woodlands, and natural and manmade environments.

All major physical development projects should be subjected to EIAs before implementation. Impose buffer zones along major river corridors and areas of outstanding natural beauty (ecological systems). Coordinate refuse dump infill reclamation and waste recycling. Safeguard and enhance the existing high-quality state of the environment within the planning area.

6.8.7 LOCAL DEVELOPMENT PLAN NO. 36

In line with the provisions of the Regional, Town, and Country Planning Act as well as the Operative Harare Combination Master Plan, the aim of the Local Development Plan is: To examine the nature of the land uses and attendant human/economic activities in the area *vis a vis* statutory planning frameworks governing the area (the Rural South Western Section and the Harare Combination Master Plan) with the following objectives in mind:

  i. To identify potentials and constraints for future growth and expansion
 ii. To assess the suitability of the area for future growth in terms of infrastructure, terrain, characteristics and zoning status
iii. To make appropriate policy recommendations that will guide development in the area for the next five to ten years

6.8.8 THE WATER ACT [CHAPTER 20:24]

The Water Act [*Chapter 20:24*] that repealed the previous one of 1976 [*Chapter 20:22*], became effective in January 2000. There are parts that are mainly concerned with the protection of the environment and hence are relevant to the operation of the smelting plant project. Part IV of the Act is concerned with the control of water pollution and the protection of the environment. In Sections 67–71 of the Act, provision is made for ensuring that water resources management is consistent with the broader national environmental approaches. The discharge of effluent or waste water into any water body is regulated by permits, which are issued with conditions in accordance to prescribed standards, and for which fees and fines are payable. The discharge of effluent or wastewater into water bodies is regulated by permits to which conditions will be attached, subject to prescribed standards, and for which fees are payable. These standards are set in SI 274 of 2000, Water (Waste and Effluent Disposal) Regulations. The classification is based on the quality of the effluent and environmental risk as submitted by the applicant and assessed by the programmable control unit (PCU). The waste disposal route can be classified as Blue, Green, Yellow, or Red according to guideline decision tables, and fees are calculated on the basis of this classification. Red is the worst scenario and blue is the most desirable scenario. For effluent in the blue classification, the water quality of the receiving body is taken into consideration. Receiving water can be classified as "normal" or "sensitive". The different color codes (see Table 6.1) have different charges such as a monitoring charge, an environmental charge, and a penalty charge.

If the permit holder does not submit information on effluent quality and quantity then the fees will be based on the previous quarter's data plus 25% or the inspector's estimate, whichever is higher.

6.8.9 PUBLIC HEALTH ACT [CHAPTER 15:09]

The Public Health Act makes provision for public health. In environmental terms, it prohibits or regulates activities that are likely to pollute streams, which in turn

TABLE 6.1
Wastewater Categorization

Classification	Risk	Reasons for Classification
Blue	Safe	Complies with blue standards
Green	Low hazard	Meets green standards not blue
Yellow	Medium hazard	Meets yellow standards but green not met
Red	High hazard	Meets red standards but yellow not met

Source: SI 274 of 2000, Water (Waste and Effluent Disposal)

become a nuisance or danger to public health. Part IV of the Act defines a nuisance as premises that promote the spread of infectious diseases, pools of water that may serve as breeding places for mosquitoes, polluted domestic water, and accumulation of refuse. The provisions of the Public Health Act must therefore be complied with throughout the operation of the smelting plant.

6.8.10 NATIONAL MUSEUMS AND MONUMENTS ACT [CHAPTER 25:11]

The Act protects all areas of historical, architectural, archaeological, and paleontological value. Such sites cannot be altered, excavated, or damaged and material on them cannot be removed without the consent of the Executive Director of the National Museums and Monuments of Zimbabwe (NMMZ).

6.8.11 FACTORIES AND WORKS ACT [CHAPTER 283 OF 1996]

The Act deals with registration and control of factories. No premises should be used as a factory unless registered. All incidents should be recorded and the relevant inspectors informed. The Act also addresses precautions against accidents to building workers. This would be relevant during both the construction and operational phases of the project. The processing plant to be established will have a factory setting and thus must be in compliance with provisions of the Act.

6.8.12 ELECTRICITY ACT [CHAPTER 13:05]

Part IV deals with the acquisition of land for power transmission and distribution, way leaves over land as well as tree and buildings interfering with transmission lines. The Act also deals with tariff, licensing, and accidents that are relevant to the residents and local authority alike. There might be need to erect small stretches of power lines and substations to supply the smelting plant.

6.8.13 EQUATOR PRINCIPLES

The principles are a set of requirements that are employed by Equator Principles Financial Institutions (EPFI) in order to ensure that the projects they finance are developed in a manner that is socially responsible and reflects sound environmental management practices. The Equator Principles are a set of voluntary guidelines for managing environmental and social issues. By doing so, negative impacts on project-affected ecosystems and communities can be avoided where possible; if these impacts are unavoidable, they can be reduced, mitigated, and/or compensated for appropriately. These principles were adopted in June 2003 by ten international commercial banks and as of June 2006, 41 EPFIs have adopted these principles, representing approximately 80 percent of global project financial Institutions. As such, the smelting plant may comply with the 2006 Equator Principles:

    i. Review and categorization
   ii. Social and environmental assessment
  iii. Applicable social and environmental standards
  iv. Action plan and management system
   v. Consultation and disclosure
  vi. Grievance mechanism
 vii. Independent review
viii. Covenants
  ix. Independent monitoring and reporting

6.8.14   WORLD BANK POLLUTION PREVENTION AND ABATEMENT HANDBOOK

The Pollution Prevention and Abatement Handbook will also be used during the EIA process. It describes pollution prevention and abatement measures and emission levels. Alternatively, use of country legislation and conditions may be used to recommend alternative emission levels and approaches to pollution prevention and abatement for the project.

Most international financial institutions and banks have introduced guidelines that compel project proponents to undertake an EIA. These guidelines are usually based on the World Bank Guidelines for Environmental Assessment (World Bank Operational Directive 4.01 – January 1999). Because these guidelines are considered to be the international benchmark for environmental assessment, an EIA complying with the World Bank Guidelines will satisfy most financial institutions. The World Bank EIA Guidelines call for screening of proposed projects to determine the type and extent of assessment required. A proposed project is classified as Category A if it is likely to have significant adverse environmental impacts that are sensitive, diverse, or unprecedented. These impacts may affect an area broader than the sites or facilities subject to physical works. Category A projects require a full EIA to examine the project's potential negative and positive environmental impacts and recommend measures needed to prevent, minimize, mitigate, or compensate for adverse impacts and improve environmental performance. A proposed project is classified as Category B if its potential adverse environmental impacts on human populations or the environment are localized and less adverse than those of Category A. Projects in category B do not require a full EIA but require environmental analysis. A proposed project is classified as Category C if it is likely to have minimal or no adverse environmental impacts. Beyond screening, no further action is required for a Category

C project. With respect to the foregoing, according to the World Bank guidelines, this infrastructure project is a Category A project and requires a full EIA and EMP.

6.8.15   Social Scan

In conducting an EIA, public involvement is essential to the process so that stakeholders' concerns are addressed. It is now generally agreed that if a development is in conflict with significant sections of the local community, this results in difficulties with regulators, generates negative publicity, and makes it more difficult to have a project approved. It is against this background that a social scan has been conducted for the smelting operation to be undertaken.

6.8.15.1   Objectives of the Public Consultation

The overarching aim of the social impact assessment is to determine whether it is socially feasible to establish the smelting operation in this location. To that end, a public consultation exercise is carried out to: evaluate socioeconomic impacts that will arise from the smelting activity and to devise ways of mitigating or enhancing them; and gain acceptance of the project by all stakeholders in order to ensure its success. This will be achieved by incorporating the concerns of the stakeholders into the project planning phase and ensuring that they are addressed.

6.8.15.2   Population of the Study Area

The Willowvale industrial area is made up of processing and manufacturing industries that offers various products and services. It is situated in the southwestern section of Harare and hosts a workforce of about 200 000 people on average. So, although it is not in the center of Harare, it is, in its own right, a robust commercial area of the city.

6.8.15.3   Social Setting

Willowvale is flanked by high-density suburbs, such as Glen View, Budiriro, Kambuzuma, Highfield, and Southerton. Most of the people who are employed in the Willowvale area reside in the aforementioned suburbs.

6.8.15.4   Land Use

Most of Willowvale is made up of industrial sites, such as manufacturers of textiles, steel, and different components. Every site has its own offices and a few infrastructures. However, most of the industries are heavily equipped with heavy machinery that is mounted into the ground or constructed on concrete slabs. In some of the industrial yards, vehicles are serviced and some of the industrial buildings are used as warehouses. There are small business units that also operate in the area, offering various services.

6.9   APPROACH AND METHODOLOGY

Secondary data and field work were used in the assessment to evaluate and predict the likely socioeconomic impacts of the proposed smelting project. A selection was made of the best research methods that will identify the major issues at stake and give the most accurate results prediction. Mitigation measures will then be recommended for the negative impacts while enhancement measures will be proposed for the positive impacts.

Due to the planned setup of properties, industrial area-based respondents for the study were known and purposely selected. Respondents for the study included government departments, company owners, residents, and employees from neighboring companies. The study targeted adults between the ages of 16 and 70 and made use of a structured questionnaire technique in order to solicit the required information. The questionnaires were conducted in face-to-face interviews in order to probe deeper on issues that required more clarity. The concerns raised were documented and taken into account in the final production of the EIA report. In conclusion, he said that there could be negative impacts on the bakery, such as traffic congestion caused by the transportation of raw materials and finished products.

6.10   ANALYSIS OF ENVIRONMENTAL ISSUES

It is indeed clear that the smelting project is being supported in the industrial area by various interested and affected parties (I and APs) because the national unemployment rate is high, and the emergence of such operations can create employment opportunities. There is actually a need for multiplication of such industries. However, the collection of scrap, storage, smelting, and distribution has many adverse impacts on the community and other industrial operations that need to be dealt with effectively. For example, collection and storage needs to be managed so that the visual setting of the place remains acceptable to the human eye. A positive impact of scrap metal collection is the reduction in the exploitation of virgin land which contains metal deposits.

Air pollution is one of the issues associated with smelters due to their high operating temperatures. The best way to deal with the emissions that are released into the atmosphere is through the use of high stacks (>50 m) and fast-blowing fans. Smoke can be dealt with by scrubbers which reduce the sulfur in smoke. It's clear according to environmental legislation that minimal air emissions should be released high into the atmosphere where dilution will effectively take place. Although there is a high potential of negative impacts, economic benefits of scrap metal smelting should not be ignored.

The smelter should benefit the locals as far as resource utilization is concerned. The industrial area and the community should benefit from the smelting initiative, thus improving the social status of the people. The very people in the industrial area must benefit from employment when the smelting project commences, but should not compromise their health. In conclusion, job opportunities will sprout for both unskilled and skilled workers in the area, thus providing survival options to the people. Consultations reveal that many in the local community are concerned that other companies fail to mitigate air pollution caused by their operations. In light of this, strict enforcement of the environmental regulations should be ensured. The use of equipment is to be considered seriously to the benefit of the entire surrounding communities.

Where possible, the cost of mitigating air pollution should be covered under the "polluter pays principle". The emissions usually affect the health of individuals as the emissions begin to affect the respiratory tracts of the employees in the area.

Therefore, the company should provide respirators to the employees to reduce the potential of inhaling large quantities of smoke. Negative impacts from rare disasters like earthquakes should be put into perspective. The smelter shall be constructed on stable ground to reduce potential dangers associated with instability.

The proponent promised that equipment to be used will be of high technical quality to such an extent that the fumes will be monitored and reduced in a significant way. This is made possible by the use of stacks and fans so that all the smoke will be projected high into the sky. While most I & APs desire to maintain a friendly relationship with their smelter neighbor, there is concern about the potential impact that pollutants might have on their own products. An agreement should be reached with other companies whereby the mitigations outlined in the EMP are adhered to so as to minimize the spread of pollutants to other premises.

Since smelters release smoke, it is also important for the company to constantly monitor its effects on neighboring properties (Fukubayashi et al., 1974). Whereas it is impossible to ignore the issues and impacts of local and global nature, it is equally important to recognize the economic and social value of such projects as this scrap metal smelter. Sustainable methods and techniques, such as green buildings and equipment can be adopted to reduce emissions. Emissions will be appropriately mitigated in every possible way. Transportation will have to be appropriately managed.

Continuous medical examinations should be regularly conducted amongst the employees so as to reduce health risks. The health of the employees could be challenged by emissions and radiation if they are not mitigated. It is critical for the smelter company to manage these risks in every way possible through pollution mitigation and the use of protective equipment by the employees in the plant.

6.11   ENVIRONMENTAL IMPACT ASSESSMENT

The section identifies environmental impacts likely to be caused by the operations of the smelter during the entire life cycle of the project from planning, designing, operational, and decommissioning stages. These impacts are analyzed and mitigatory measures proposed. The proponent of the project considers these proposals, and project designers are guided accordingly. The EMP is constructed on the basis of this EIA report. The company's planned maintenance program also considers the areas identified to be likely sources of pollution.

6.11.1   IDENTIFIED ENVIRONMENTAL CONCERNS

Identification of environmental concerns requires that a checklist be constructed listing all environmental issues likely to be affected by the project. As a result of this identification, a cause and effect matrix was drawn up that encompassed the whole of the project cycle. Major environmental concerns in the area of the project were identified and are listed below:

  i. Aesthetic standards, visual, and landscape setting
 ii. Sociocultural and socioeconomic issues
iii. Water quality

   iv. Land use conflicts
    v. Environmental constraints
   vi. Flora and fauna
  vii. Public health and safety
 viii. Soils
   ix. Air quality including noise, dust, and vibration

In any EIA report, impact analysis considers the main areas that include:

Nature of Impacts

A project impacts the environment in which it is to be situated both negatively and positively. The beneficial impacts should outweigh the detrimental effects for a project to be feasible. There must be a plan to mitigate against the adverse effects of the project.

Spatial Scale

   Local (the site and immediate surroundings)
   Regional (provincial scale)
   National
   International

Duration

   Short-term (0–5 years)
   Medium-term (5–15 years)
   Long-term (for the lifetime of the activity)

Significance

Low, where the project will not have an impact on the decision to go ahead with the development

   Medium, where the project may have an impact on the decision to go ahead with the development, unless mitigated

   High, where the project will influence the decision to go ahead with development, regardless of possible mitigation

6.11.2   Impacts during the Planning Phase

Stakeholder Involvement

For any developmental project to be successfully implemented, stakeholders' buy-in into the project cannot be overemphasized. People need to understand the impact of the project on their lives

Significance of Impact

   Negative [-]
   Local
   Medium-term
   Medium

Mitigation

Stakeholders should be involved in the project life. The company establishing the new project in the area should practice good corporate governance in its operations and social corporate responsibility through activities that benefit the community.

6.11.3   REGULATORY REQUIREMENTS

The several environmental laws that have been promulgated are not meant to punish investors but rather are meant to encourage sustainable development. It is therefore imperative for the project proponents to adhere to all statutory requirements before the commencement of the smelter establishment.

Significance of Impact

> Positive [+]
> Local
> Long-term
> Low

Mitigation

Before the commencement of the project, there is need to satisfy all the lawful requirements. The lawful requirements include all the permits from the statutory bodies like EMA, local authorities, and any other institution that may have a stake in the development of the smelter.

6.11.4   SERVICE INFRASTRUCTURE

The introduction of a new project in an area may mean stress on the available resources such as water and roads.

Significance of Impact

> Negative [-]
> Local
> Long-term
> Low

Mitigation

Service providers, like water and electricity suppliers, should be made aware of the requirements of the new project. This enables the service providers to make available enough of the utilities so that the new development does not impede service to other users.

6.11.5   WILDLIFE HABITAT

Bush clearance in the project area disturbs wildlife habitat. The habitat may be that small ant heap which is home to a colony of thousands of ants.

Significance of Impact

Negative [-]
Local
Short-term
Low

Mitigation

Vegetation clearance should be confined to only the areas where buildings are to be erected to avoid disturbing wildlife habitat.

6.11.6 EROSION

Soil erosion is directly related to increased runoff and rainfall intensity. Built-up areas generally inhibit infiltration due to the presence of roofs, concrete slabs, and tarred driveways, thereby increasing runoff.

Significance of Impact

Negative [-]
Local
Short-term
Low

Mitigation

There is little evidence of erosion in the area, as the general aspect of the area does not encourage high erosion rates. There is need to plant lawn in all open spaces at the site. There is also need for the civil works to put concrete-lined storm drains in place that direct all runoff water to the HCC main storm drainage system.

6.12 IMPACTS DURING THE SMELTING OPERATIONS

The proper smelting operation brings with it a lot of environmental challenges as activities from delivery of the scrap up to the storing of the ingots begin to take place. Environmental issues likely to be encountered include:

   i. Soil disturbances
  ii. Water and air pollution
 iii. Noise and vibration
  iv. Vegetation loss
   v. Wildlife habitat disturbance
  vi. Health and safety
 vii. Disturbances to traditional cultural sites and artefacts

6.12.1 VEGETATION LOSS

Vegetation will be cleared on the area where buildings are to be erected. The vegetation cleared shall mainly be grass as most trees were already cleared by the urban agriculture activities that were taking place in the area prior to the advent of the smelter plant.

Significance of Impact

    Negative [-]
    Local
    Short-term
    Low

Mitigation

Identification of species that are found in the affected area needs to be done so that a replacement process is then conducted after the construction works at the site have been completed. Clearing of vegetation should only be limited to the area where buildings are to be constructed.

6.12.2 DISTURBANCE OF ARCHAEOLOGICAL AND CULTURAL SITES

Archaeological and traditional sites may be present in the project site area, so archaeologists should scour the area for evidence of the presence of any arte-facts and shrines. Marginal lands on wetlands were and are still used for infant burials.

Significance of Impact

    Negative [-]
    Local
    Short-term
    Low

Mitigation

It may be difficult to assertively link any artefact found in the area to a certain cul-ture, since Harare is now a diverse and fully urbanized city. Museum curators are helpful in the identification of traditional artefacts and practices. Any exhumations that need to be done, according to the Act, are the responsibility of Home Affairs Ministry.

6.12.3 LOSS OF ARABLE LAND

The area on which the project is to be established was previously being used for subsistence urban agriculture. The agriculture, however, is not legal in the sense that the cultivators are not bona fide owners of the land.

Significance of Impact

    Negative [-]
    Local
    Short-term
    Low

Mitigation

The subsistence farmers' loss of agricultural land is inevitable, as the Council sells land to prospective investors. The City Council shall be consulted on how best the farmers may be assisted with alternative land if available.

6.12.4   WILDLIFE HABITAT DISTURBANCE

Wildlife encompasses all the living organisms that live in the wild from the microorganisms in the soil up to the large mammals that may have habitat in the project site. The disturbance to the natural habitat of the wildlife may cause forced migration to other places where the species may be predated, or adaptation to the new environment may not be possible and result in the death of the species.

Significance of Impact

Negative [-]
Local
Short-term
Low

Mitigation

It may be impossible to preserve all habitats, but, where possible, great care must be taken to protect the habitats. Replanting of the original plant species may help in that these plants may have had a dual role of being both a food source and home to some species.

6.12.5   LAND OWNERSHIP WRANGLES

Land use conflicts emerge if the activities on the project site affect activities in the adjacent sites. This conflict may arise from emissions, noise, drainage, and fouling of air and many other pollution-related issues that may arise as a direct result of the activities on the smelting project site. Land ownership may also be a cause of conflict as there may be double allocation of the same land by Council.

Significance of Impact

Negative [-]
Local
Short-term
Low

Mitigation

The adjacent properties should be made aware of the activities of the project to be established. The public consultation exercise should be as transparent as possible and must capture the concerns of the adjacent operations so that project designers take into consideration the issues raised to incorporate into the project design. A deeds search should be done in order to avoid ownership wrangles that may be costly.

6.12.6   DISTURBANCE OF WATER BODIES

The project is likely to cause little effect to the infiltration rate to the ground below the project area and also little effect to the vlei-like area as the area, covered by buildings, is small compared to the whole property. The stream in the area is an ephemeral one, and, as such, there is little effect to it except for increased runoff from the roof and driveways during the rainy season..

Significance of Impact
Negative [-]
Local
Medium-term
Low

Mitigation
Efforts must be taken to avoid erosion of the soil which may eventually be deposited into the stream in the project area. Any excavated soil should be disposed of in a manner that it will not block free flow of water in the natural channels.

6.12.7   LOWERING OF GROUNDWATER LEVELS

Construction works, as previously, stated reduce infiltration, thereby affecting groundwater recharge. When there is reduced recharge, the water table sinks, or lowers, therefore affecting plant life, as some plants do not have deep roots to reach the water table. Larger plants may not be able to draw the water from hygroscopic soil particles.

Significance of Impact
Negative [-]
Local
Short-term
Low

Mitigation
Planting of trees and grass reduces the speed of runoff and also mitigates raindrop impact which compacts the soil. The trees and grass therefore cause the water to sink into the ground.

6.12.8   EFFECTS ON WATER QUALITY (GROUND AND SURFACE WATER)

The project is unlikely to cause major changes in water quality due to the fact that no chemicals are going to be used in the smelting process. The source of water pollution may be from the rust on the scrap metal which is in the stackyard. The other possible source would be oil leaks from the haulage trucks. The coal stack can also contribute to water quality problems. Both leaching and runoff may cause pollution of both ground and surface water, respectively.

Significance of Impact

Negative [-]
Local
Medium-term
Low

Mitigation

The stackyard concrete slab should have a concrete lined trench around it and some trap sumps put in place to trap oils and rust. Periodic cleaning of the traps should be done to remove said trapped oils and rust. The coal pit should be concrete-lined so as to stop any leaching of the ash. Burned coal should be trucked off to City Council landfills or dumpsites or it can be donated to be used in brick molding or fill material at construction sites.

6.12.9 SCRAP METAL STORAGE

The scrap metal yard may cause land pollution if the scrap metal is not stored properly. Rodents may find new homes in the scrapyard, thereby affecting adjacent businesses. The scrap metal also can be a source of injury to the workers if unstable stacks fall on workers.

Significance of Impact

Negative [-]
Local
Long-term
Low

Mitigation

The scrap metal should have a height of not more than 2 m so as to prevent falling of scrap metal from heights. Indeed, the scrap metal should be stacked by skilled individuals. Compaction of the scrap by hydraulic presses may also help in the orderly storage of the scrap.

6.12.10 WASTE HANDLING

Waste management poses a challenge when there is no safety, health, and environment (SHE) personnel and if the smelter does not have an environmental management system (EMS) in place. Failure to handle environmental issues results in pollution of the environment.

Significance of Impact

Negative [-]
Local
Short-term
Low

Mitigation

The company should implement an EMS so that all environmental concerns are addressed. Regular environmental audits should be conducted to ensure adherence to environmental laws.

6.12.11   Waste Management on Site

Waste management poses a great challenge to most project developers, in that they tend to dump waste just outside the site. This is mainly due to the fact that most developers do not have refuse removal trucks to take waste to dump sites. The smelter does not produce effluent, but it does produce smoke from the smelting process.

Significance of Impact

> Negative [-]
> Local
> Short-term
> Low

Mitigation

During the construction stage, the smelting company should have refuse trucks to transport excavated soils for disposal at landfills in the City Council-approved sites. Mobile bins should be placed on-site to collect all refuse.

6.12.12   Access Roads

The smelting plant is close to major roads in the area. There shall be disturbance to soil as it is removed during construction of the roads. Gravel shall be taken from other places, thereby creating pits in those areas. The access road will affect infiltration rate as already alluded to in this document.

Significance of Impact

> Negative [-]
> Local
> Short-term
> Low

Mitigation

The trucks that bring scrap metal to the smelting plant should avoid using the busy roads during the peak traffic hours of 7:00–9:00 am and 4:30-6:00 pm. Any deliveries should be outside these times to avoid congestion. The access road to the smelter should be tarred to withstand the weight of the delivery trucks.

6.12.13   Safety, Health, and Environment

Smelting plants are potential fire hazards and there are high risks of burns from the molten iron and the iron ingots that have not cooled (Bale et al., 2002). Workers and visitors to the plant who are not properly protected by protective clothing or fire suits run a high risk of fire-related injuries. These are the most common type of injuries in smelting plants. Industrial injuries also arise from use of machines and have led to death and decapitation of workers. Other injuries occur in the scrapyard and from electric shocks.

Significance of Impact

Negative [-]
Local
Short-term
Low

Mitigation

The company should ensure that appropriate protective clothing is purchased and safety rules emphasized on site. This can be enhanced by the use of relevant safety procedures.

6.12.14   EMPLOYMENT OPPORTUNITIES

The smelting plant shall create employment opportunities for 45 people in the initial stage. There are prospects of expansion of the operation if the demand for the smelter products rises.

Significance of Impact

Positive [+]
Local
Short-term
Low

Enhancement

The smelter company employment policy should deliberately consider employing people who live closer to the factory especially the unskilled laborers. This has the advantage that since the company will operate 24 hours a day. This will also reduce the burden to the company on transport costs, as the workers live close to the factory.

6.12.15   HIV/AIDS ISSUES AND OTHER SOCIAL VICES

Social vices such as prostitution, muggings, and HIV/AIDS may increase as workers would have disposable income.

Significance of Impact

Negative [-]
Local
Medium-term
Low

Mitigation

Workplace peer education on HIV/AIDS should be encouraged and condoms should be placed at points, especially the toilets, where the workers can easily access them.

6.12.16   NOISE POLLUTION

Potential Impact

Noise is often regarded as a nuisance rather than a widespread occupational hazard. The three major categories of noise source are:

    i. Internal fixed
   ii. Mobile plant
  iii. External transport movements

Fixed plant machineries are associated with the day to day smelting and generate noise. Mobile plant machinery such as transport movements comprise operations like supply of raw materials, dispatch of finished products, handling and disposal of waste by the road.

Significance of Impact

    Negative [-]
    Local
    Long-term
    Medium

Mitigation

Planned maintenance is vital in industry as moving machinery parts are continually lubricated to reduce friction which is a source of noise. There is also need to look at the depreciation rate of the plant equipment and vehicles so that timely replacements are acquired, firstly, to reduce operational costs and, secondly, to reduce noise levels.

6.13   TRANSPORT IMPACTS

6.13.1   ACCIDENT RISKS

Potential Impacts

High volume of traffic increases the chances of accidents on the roads. Most accidents are caused by bad roads, bad road signs, and the unroadworthy vehicles that ply the roads.

Significance of Impact

    Negative [-]
    Local
    Long-term
    Medium

Mitigation

There should clear road markings on the roads and all intersections should be controlled by traffic lights or traffic police.

6.13.2   PEDESTRIAN SAFETY

Potential Impacts

The busy roads in the area may pose a serious threat to pedestrians as there is great danger of loss of life and injury due to traffic.

Significance of Impact

> Negative [-]
> Local
> Medium-term
> Medium

Mitigation

Special pedestrian crossing facilities should be provided at strategic locations on the roads so that pedestrians can cross safely. Road awareness campaigns should be coordinated on a regular basis or road safety posters should be hung in the workers' canteen.

6.13.3   ROAD MAINTENANCE

Potential impacts

Especially at intersections, roads give way to the pressure exerted by heavy trucks to the point that potholes develop at these intersections.

Significance of Impact

> Negative [-]
> Local
> Long-term
> Low

Mitigation

The intersections should have compressed interlocking concrete bricks that can withstand the weight of turning trucks. Trucks should not carry overloads, as this has an effect on the roads.

6.13.4   VEHICLE EMISSIONS

Potential Impact

The heavy vehicles have a huge polluting potential in the form of emissions as a result of poor maintenance.

Significance of Impact

    Negative [-]
    Local
    Medium-term
    Low

Mitigation

All vehicles must be regularly serviced and kept in a good operating condition to minimize toxicity of emissions in the atmosphere.

6.13.5   NOISE POLLUTION

Potential Impact

Huge haulage trucks cause a lot of noise from their big engines which are a nuisance to persons and institutions within the vicinity of the operations. The noise is a source of vibration especially to buildings with glass compartments or poorly sound proofed buildings.

Significance of Impact

    Negative [-]
    Local
    Medium-term
    Low

Mitigation

Haulage trucks may produce significant amounts of noise and vibration. To minimize this effect the trucks must be regularly serviced and maintained. The trucks also must travel at low speeds to reduce the noise especially when trying to stop at intersections, where, at times, the driver may be forced to use the exhaust braking system.

6.13.6   DUST

Potential Impact

The movement of traffic in the area is likely to cause dust. This is common when the roads are narrow or at road bends.

Significance of Impact

    Negative [-]
    Local
    Medium-term
    Low

Mitigation measures

All roads should be tarred and they must be wide enough in accordance with City Council bylaws on road size for the service road to the plant. Quarry stones should be put in those areas in the plant yard where there is no concrete to reduce dust emissions.

6.13.7 OCCUPATIONAL HEALTH AND SAFETY

Potential Impact

The plant may be a source of injury to workers when safety precautions are not observed. There are safety precautions that must be observed when operating machines and general safety precautions of the work place environment. Machines that are not adequately serviced are a danger to the workers.

Significance of Impact

> Negative [-]
> Local
> Medium-term
> Low

Mitigation

A code of conduct should be put in place to regulate behavior in the plant area. All the employees should be required to read and sign the code of conduct, having seen and read it together with the safety rules of the company. Planned maintenance should be performed regularly in order to ensure safety of machinery and the buildings in general. There must be a proper fleet management plan for the vehicles to ensure that they are roadworthy.

6.13.8 FUEL SPILLS

Potential Impacts

There is a high risk of groundwater contamination as well as fire outbreaks resulting from oil and fuel leaks from vehicles.

Significance of Impact

> Negative [-]
> Local
> Short-term
> Low

Mitigation Measures

The areas likely to experience oil spills are the scrap metal stackyard and the refuelling area. The stackyard shall put into place a concrete slab, an oil trap trench, and sump.

6.13.9 FIRE

Potential Impacts

Fires are likely to be caused when flammable items come in contact with heat. The molten iron from the furnace can cause fires when paper, plastic, cloth, and grass come into contact with it.

Significance of Impact

    Negative [-]
    Local
    Short-term
    Low

Mitigation Measures

There is a potential risk of fire outbreaks at fuel storage points. Signs should be put in place to warn of the potential danger. A fire guard must also be put around the premises to avoid the spread of fire onto neighboring properties or from outside into the factory. There must be fire-fighting water hydrants at strategic places especially those areas with a high fire potential. Fire escape routes must be clearly labelled and an assembly point clearly designated in the factory area. Fire extinguishers and sand buckets must also be placed in visible and accessible areas.

6.13.10 Wastes

Potential Impacts

There is a risk of groundwater contamination from the disposal of wastes associated with operations of loading trucks, namely used oils, grease, and fuel.

Significance of Impact

    Negative [-]
    Local
    Short-term
    Low

Mitigation Measures

Trucks should have regular servicing, and all spills of oil, grease, and fuel must be cleaned immediately.

6.14 IMPACTS AFTER ESTABLISHMENT OF SMELTING PLANT

6.14.1 Degradation of Surface Waters

Surface water degradation results from introduction of pollutants from industrial or domestic activities. The pollutants may be solid or liquid. Potential pollution may come from the rust in the scrap metal stackyard and petroleum oil spills.

Significance of Impact

    Negative [-]
    Local
    Short-term
    Low

Mitigation
The company should ensure that no scrap metal stays long in the stackyard before processing. The badly rusted scrap metal should be processed first. Oil spills must be attended to with urgency.

6.14.2 MODIFICATION OF VEGETATION

The clearing of vegetation in area destroys the native species in that area and new vegetation is introduced in the form of decorative shrubs and grasses.

Significance of Impact
> Negative [-]
> Local
> Short-term
> Low

Mitigation
It is essential to undertake prompt reclamation of disturbed areas with native species after the smelting process.

6.14.3 VEGETATION LOSS

Vegetation clearing to make space for construction of buildings causes the loss of vegetation in the area.

Significance of Impact
> Negative [-]
> Local
> Short-term
> Low

Mitigation
Efforts should be made to re-vegetate those areas that have been cleared, and planting of the original species should be encouraged.

6.14.4 CORPORATE SOCIAL RESPONSIBILITY

Most corporate institutions do not take environmental issues seriously, as their main objective is to make a profit. Unfortunately, this happens at the expense of the environment in which they operate. They only react when EMA bills them for violating environmental laws. A good environmental work place plan is good business.

Significance of Impact
> Positive [+]
> Regional

Long-term
Low

Mitigation

The company should, through its public relations department, engage in community work that helps to address environmental issues in the area, like providing waste/recycling bins or sponsoring an environmental club at a local school. Other social responsibility activities may include sponsoring youth sporting activities in the area to reduce the prevalence of social vices such as drug addiction, crime, etc.

6.14.5 FEEDBACK TO STAKEHOLDERS

Once the smelter is fully functional, new environmental challenges may arise which were not anticipated during the assessment phase. This can become a problem, especially if communication between the company and the immediate community has waned over time, so much that there is no direct link between the two parties.

Significance of Impact

Negative [-]
Local
Short-term
Low

Mitigation

The company must take initiative at all times to inform the public about any issues that may concern them or affect them. Billboards, radio, and television are effective means of disseminating information and warnings to the public.

6.15 SUMMARY OF THE ENVIRONMENTAL IMPACTS

While most of the potential environmental impacts are negative, the significance of most impacts is medium and can be mitigated sufficiently. Table 6.2 summarizes the environmental impacts.

6.16 ENVIRONMENTAL MANAGEMENT PLAN

An EMP was formulated based on the impact analysis exercise to implement the mitigation measures identified and enhance the positive impacts of the project. An EMP is a tool that is used in the mitigation process in the life cycle of a project. EMP is principally an integrated effort of utilization, planning, maintenance, supervision, control, recovery, and development of the environment. The success of the plan very much depends on the method and techniques of management to be implemented. The EMP must be understood by all in order to gain the trust and cooperation of the entire workforce.

TABLE 6.2
Summary of Environmental Impacts

Impact	Before Smelting Operations	During Smelting with Mitigation	After Smelting with Mitigation	Overall Impact with Mitigation
Health problems through increased dust concentrations	None	Low	Low	Low
Destruction and disturbance of indigenous vegetation and habitats	None	Medium-High	Medium-High	Medium-High
Destruction of faunal habitat	None	Medium–Low	Medium-Low	Medium
Dust	None	Medium	Medium-Low	Low
Increased ambient noise levels affecting surrounding community	None	Medium-Low	Medium-Low	Medium
Loss of scenic integrity due to alteration of visual character	None	Medium-Low	Medium-Low	Medium
Increase in secure jobs and incomes through project recruitment	None	Medium-Low	Medium-Low	Medium
Increased business opportunity through procurement of goods and services	None	Medium-Low	Medium-Low	Medium
Local improvements to road, power, and water infrastructure, with benefits to proximate communities	None	Medium-Low	Medium-Low	Medium

(Continued)

TABLE 6.2 (CONTINUED)
Summary of Environmental Impacts

Impact	Before Smelting Operations	During Smelting with Mitigation	After Smelting with Mitigation	Overall Impact with Mitigation
Increased local government service delivery capacity due to establishment of smelter plant	None	Medium-Low	Medium-Low	Medium
Reduced pressure on land resources due to secure incomes and livelihoods among employees	None	Medium-Low	Medium-Low	Low
Disturbance of wild animals	None	Medium-Low	Medium-Low	Low
Increased local risk of HIV/AIDS infection with influx of workers and opportunity seekers	None	Medium-Low	Medium-Low	Medium-Low
Access roads will cause compaction or erosion to the surface	None	Medium-Low	Medium-Low	Medium
Reduced water availability for domestic use	None	Medium-Low	Medium-Low	Medium
Waste handling and waste management on site	None	Medium-Low	Medium-Low	Medium-High
Oil and fuel leaks from vehicles	None	Medium-Low	Medium-Low	Medium
Changes in surface and groundwater quality due to on-site activities	None	Medium-Low	Medium-Low	Medium
Lowering of groundwater levels	None	Medium-Low	Medium-Low	Medium
Fire	None	Medium-Low	Medium-Low	Medium-High
Smoke	None	Medium-Low	Medium-Low	Medium

TABLE 6.3
EMP for the Planning Stage of Smelter Establishment

Issue	Management of Issue	Implementing Agency	Monitoring Agency
Stakeholder Involvement	People tend to be hostile to a project if they are not aware of it, and this may inhibit development. It is essential that their concerns are raised before project initiation to avoid unnecessary project delays. Cooperative and open working relations should be established early with communities and maintained throughout the life of the project.	Consultant Company	EMA HCC
Regulatory Requirements	It is essential to fulfil all regulatory requirements before commencement of any work on the site to not only protect the reputations of parties concerned but also to maintain project sustainability.	Company Consultant	EMA HCC
Service Infrastructure	It is essential to ascertain the requirements for service infrastructure required for the proposed smelting project so as to ensure that the existing resources are not strained to the detriment of the communities in the area.	Company	HCC
Wildlife Habitat	The proposed smelting project will result in reduced reproduction in the population from habitat modification or loss. There is a need for the avoidance or restriction of disturbance of significant wildlife habitats. Vegetation clearing should be minimized as much as possible in order to retain habitats for various wildlife species.	Company	EMA HCC
Erosion	Vegetation that has naturally grown in and around the area should not be removed. At present there is no evidence of major erosion.	Company	EMA HCC
Cleared Linear Facilities (access roads)	It is essential to utilize existing access roads and minimize clearing.	Company	EMA HCC
Cultural Issues	It is essential to work with local authorities such as the Chief to ensure clarity on remains.	Company Consultant	NMMZ

6.16.1 EMP IMPLEMENTATION PLAN

This section deals with the EMP as regards planning, operational, and post-operational activities. The management of all possible impacts is of paramount importance as failure to manage may result in penalties being imposed by the relevant state enforcement agencies. It may also result in injury, death, and, above all, pollution of

TABLE 6.4
EMP for the Smelter Operation Stage

Issue	Management of Issue	Implementing Agency	Monitoring Agency
Modification of Soil Profile, Vegetation, and Surface Drainages	During the plant construction, there will be need for the segregation and stockpiling of topsoil for use in reclamation	Consultant	EMA Consultant
Decreased Capacity of Local Reservoirs	There are nearby water bodies in the proposed area and these may become silted. There is a need to avoid the disturbance of the vlei. Where disturbance cannot be avoided, sediment control structures should be put in place	Consultant	EMA HCC
Wildlife Habitat	During the smelting project, there may be wildlife deaths from road traffic and surface disturbances; there is a need to emphasize driver awareness. Vegetation clearing should be minimized as much as possible in order to retain habitats for various wildlife species	Consultant	EMA HCC
Degradation of Air Quality	Smelting project may result in the generation of airborne particulates which may be a nuisance. There is a need for proper blasting to minimize airborne particulates and watering of haulage roads.	Consultant	EMA HCC
Noise and Vibration	If any changes occur, the residents should be notified. Communication is very important in maintaining a good relationship with the locals. Employees should wear ear defenders during operation.	Consultant	EMA
Employee Health and Safety	There is a need to ensure that appropriate safety and rescue equipment is available and employees are trained in its use. Contractors are asked to adhere to good operational practice. This will include dust suppression by water spraying, where necessary	Consultant Contractors	EMA HCC
Archaeological Resources	Archaeological resources discovered during excavation should be reported to the National Museums and Monuments of Zimbabwe. However, to avoid vandalism of identified artefacts these should not be publicized	Consultant	NMMZ EMA
Vegetation Loss	Efforts should be made to revegetate those areas where clearing has taken place	Company	EMA HCC

(*Continued*)

TABLE 6.4 (CONTINUED)
EMP for the Smelter Operation Stage

Issue	Management of Issue	Implementing Agency	Monitoring Agency
Legal Compliance	All regulatory framework should be complied with during the smelting project; this includes payment of levies to the local authority and submission of quarterly environmental audits	Company Consultant	EMA HCC
Waste Handling	Most of the waste produced during the construction phase includes materials that do not pose any threat of contamination of water	Company	Company
Reduced Water for Reservoirs	Recycling and treatment of water will be part of the operational philosophy of the proposed extension plan in order to reduce the need for water in the proposed weir. Please check the attached water balance sheet	Company	ZINWA
Access Roads	Traffic should be confined to specific routes and roads. Where possible, roads should be surfaced	Company	Company HCC
Road Safety	Contractors need to adhere to HCC and Ministry of Transport regulations relating to construction vehicles, road work signs, and markings	Company	HCC
Smoke	Employees should be furnished with respirators on every shift. This will ensure worker health and safe operations at all times. High-speed extraction fans with shall blow all the smoke off from inside the smelter building	Company	EMA NSSA MoHCW
Fire	The project site should be furnished with adequate fire-fighting equipment. This includes fire extinguishers, water hoses, sand buckets, and alarms	Company	EMA NSSA HCC
Heat	The employees should be furnished with insulated leather aprons that will be worn during operation	Company	EMA NSSA
Impact of Accidental Fires	A permanent firebreak around the proposed construction site should be considered before the commencement of any construction activities	Company	EMA HCC

(*Continued*)

TABLE 6.4 (CONTINUED)
EMP for the Smelter Operation Stage

Issue	Management of Issue	Implementing Agency	Monitoring Agency
Loss of Rare and Endangered Species	Discourage planting of foreign plants especially those that are classified as "invasive alien species" in the EMA Act. There are bushes of *lantana camara* nearby	Company	EMA HCC
Harmful Effects of Smelter	COMPANY should use respirators and should educate employees on metal handling	Company	EMA HCC

the environment. It is important to note that it requires total commitment of management and the workforce of the company to uphold the provisions of the EMP. The contractor is expected to have control over environmental impacts occurring within a certain defined area.

6.16.2 EMP FOR THE PLANNING STAGE

The EMP for the planning stages is shown in Table 6.3 and Table 6.4.

6.16.3 EMP FOR TRANSPORT IMPACTS

The EMP for transport impacts at the smelter plant are summarized in Table 6.5.

6.16.4 EMP FOR POST SMELTING OPERATIONS

The EMP for post-smelting operations is shown in Table 6.6 and Table 6.7.

6.17 CONCLUSION AND RECOMMENDATIONS

6.17.1 CONCLUSION

Based on the receiving environment, the potential environmental impacts due to the planning, operation, and post-smelting activities have been assessed with respect to the environmental aspects such as air quality, noise, water quality, and waste management. With the implementation of appropriate mitigation measures specified in the EMP, the smelting project will have limited and acceptable levels of adverse impacts on the environment. Adherence to this EMP will mitigate the potential negative impacts that have been outlined and make the project sustainable. The area under assessment has been previously used as an automobile service garage. The area had been sustainably used. Based on this and for the purposes of the smelting

TABLE 6.5
EMP for Transport Impacts at the Smelter Plants

Issue	Mitigation/Procedure	Implementing Agency	Monitoring Agency
Accident Risks	Traffic control signs should be clearly marked and placed where they are visible Trucks must maintain low speeds when coming in and going out of the yard to ensure that the drivers are in total control of their vehicles.	Transport Operator/ HCC	HCC ZRP
Pedestrian Safety	Special pedestrian crossing facilities should be provided at strategic locations Road awareness campaigns should be considered in built-up areas	Transport Operator/ HCC	HCC
Road Maintenance	Heavy loading of local roads by haulage trucks will increase maintenance requirements. Frequency of maintenance may need to be increased	HCC	HCC
Vehicle Exhaust Emissions	All vehicles must be regularly serviced and kept in good operating condition to minimize toxic emissions in the atmosphere	Transport Operator	EMA
Noise and Vibration	Haulage trucks may produce significant amounts of noise and vibration. To minimize this effect, the trucks must be regularly serviced and maintained	Transport Operator	Transport Operator
Dust Emissions	The vehicles going in and out of the yard kick up dust, thereby resulting in poor air quality on a macro-scale. Water spraying should be used to suppress dust that may arise due to the movement of vehicles.	Transport Operator	EMA
Occupational Health and Safety	All employees especially those working in the repair workshops must be provided with adequate personal protective clothing (PPE) to prevent injury.	Transport Operator	NSSA
Heat	All those in the vicinity of the furnace are subject to heat from the furnace because of the high operating temperature. These employees have to wear leather aprons, leather gloves, and heat-resistant goggles.	Company	EMA
Fire	There is a potential risk of fire outbreaks at fuel storage points. Signs should be put in place to warn of the potential danger.	Company	EMA
Wastes	The transport business produces quite a number of wastes through spillages, e-waste and metals	Transport Operator	EMA HCC

TABLE 6.6
EMP for Post-Smelting Operations

Issue	Management of Issue	Implementing Agency	Monitoring Agency
Modification of Vegetation	During and after the smelting project, there may be the emergence and introduction of invasive alien species. It is essential to take prompt reclamation of disturbed areas with native species.	Company	EMA HCC Consultant
Vegetation Loss	Efforts should be made to revegetate those areas where clearing has taken place.	Company	EMA HCC
Corporate Social Responsibility	A sense of social corporate responsibility should be realized with participation in community activities as well as participation in community development projects which benefit the community at large.	Company	EMA HCC
Feedback to Concerned Stakeholders	It is essential to communicate the results of the smelting project with all the concerned stake holders.	Company Consultant	EMA HCC

project, both land uses can be compatible through adherence to the EMP presented in this study. Employing engineering solutions (to improve ground stability) and the necessary mitigations can easily solve the negative impacts. There are significant positive impacts that can be associated with the proposed project.

The conclusions about the project are as follows:

  i. Reliable means exist for ensuring that the impact management measures can and will be adequately planned and implemented
 ii. Potential residual impacts on the environment from the proposed smelting project are likely to be minor
iii. The undertaking of the smelting project will not displace any people, families, or communities
 iv. The smelting activity does not significantly affect any environmentally sensitive areas
  v. The activity will use natural resources in a way that will not pre-empt the use, or potential use, of that resource for any other purpose
 vi. The planning, operation and post-smelting activities, their environmental impacts, and measures for managing them are well understood

6.17.2 Recommendations

Smelting of waste metal to bars and ingots is a recommended technology for waste management and value addition of waste metal. A comprehensive EIA must always be adopted and approved.

TABLE 6.7
Smelting Plant Overall Environmental Monitoring Plan

Issue	Mitigation	Imple- menting Agency	Moni- toring Agency	Time Frame
AIR QUALITY ISSUES				
Dust in the Site	Perform periodic water-sprinkling non-smelted lead roads.	Company	EMA	Currently being practiced
Dust at the Entrance	The entrance will be concrete paved.	Company	EMA	Permission has been sought from City of Harare
Vehicle Assembly Site	Perform periodic water-sprinkling at the assembly point	Company	EMA	Implementation to start soon after re-opening. Should be practiced regularly throughout the life of the plant
Offloading Trucks	Cover the pile or trucks with a tarp.	Company	EMA	Where possible crush stone according to orders or requirement. Immediately.
WATER QUALITY ISSUES				
Groundwater and Aquifer Contamination	A key operational design objective should be to plan the extent and depth of the workings to minimize any significant impact on groundwater resources and associated seasonal vlei features.	Company	ZINWA	Depth of the workings to be determined immediately
HEALTH AND SAFETY ISSUES				
Access control	To prevent unauthorized entry, guards need to be placed at all access points leading to the plant. In addition, barricades with warning signs may be erected at access points and should be clearly legible and bold, such as "**Warning, Smelting Plant in Operation**".	Company	EMA	Implementation should begin immediately and continue throughout the life of the plant

(Continued)

TABLE 6.7 (CONTINUED)
Smelting Plant Overall Environmental Monitoring Plan

Issue	Mitigation	Implementing Agency	Monitoring Agency	Time Frame
Effective communication	Workers and visitors should be informed about smelting operations, evacuation procedures, and location and timing of scheduled blasts.	Company	NSSA	Implementation should begin immediately and continue throughout the life of the plant
Training and Awareness	Site-specific training relative to the extent of the smelting area, non-work zone beyond the smelting area, plant guarding, access control, and clearing protocol should be made available to all new employees.	Company	NSSA	Implementation should begin immediately and continue throughout the life of the plant
OTHER OPERATIONAL ISSUES				
Noise pollution	Company should replace and select machinery and equipment for optimum utilization. Regular, planned maintenance of machinery and equipment and replacement of damaged and worn out parts should be conducted.	Company	NSSA	Implementation must start soon after opening

REFERENCES

Bale, C. W., Chartrand, P., Decterov, S. A., Eriksson, G., Hack, K., Ben Mahfoud, R., Melançon, J., Pelton, A. D. and Petersen, S. (2002). FactSage Thermochemical Software and Databases. *Calphad Journal* 62, 189–228.

Danielson, J. A. (Ed.) (1973). *Air Pollution Engineering Manual* (2nd Ed.), AP-40, U. S. Environmental Protection Agency, Research Triangle Park, NC.

Fukubayashi, H. H. and Higley, L. W. (1974). Recovery of Zinc and Lead from Brass Smelter Dust, Report of Investigation No. 7880, Bureau of Mines, U. S. Department of The Interior, Washington, DC.

Index

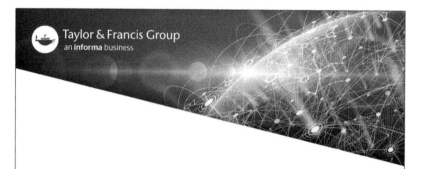

Milton Keynes UK
Ingram Content Group UK Ltd.
UKHW040101071024
449327UK00019B/725